Twenty Count

TWENTY COUNT

SECRET MATHEMATICAL SYSTEM OF THE AZTEC/MAYA

ROGER MONTGOMERY

foreward by Haleh Pourafzal

LIBRARY OF CONGRESS CATALOGING-IN-PUBLICATION DATA

Montgomery, Roger.

Twenty count : secret mathematical system of the Aztec/Maya / Roger Montgomery.

p. cm.

Includes bibliographical references.

ISBN 1-879181-26-6

1. Spiritual life. 2. Indians of Mexico–Religion–Miscellanea. 3. Mayas–Mathematics–Miscellanea. 4. Aztecs–Mathematics–Miscellanea. I. Title.

BL624.M662 1995

299'.784–dc20 95-17484

CIP

Bear & Company, Inc.
Santa Fe, NM 87504-2860

Cover Design: Lightbourne Images
Sunstone Illustration: Martin Brennan
Text Illustrations: Roger Montgomery
Text Design & Typesetting: Melinda Belter
Editing: Gerry Clow and Brandt Morgan
Printed in the United States of America by BookCrafters
1 3 5 7 9 8 6 4 2

Deepest gratitude to May Murakami,
sorceress of light and music,
for the gracious gift of listening.

CONTENTS

(illustrations preceding chapters are set in italic)

Foreword

As a Persian woman, my consciousness was inflamed in childhood by the burning poetry of Jalal-ud-Din Rumi and Omar Khayyam. I marveled at these poets' eloquence, feeling their pure yearning for freedom in the depth of my heart. I could sense in their fiery verses the catharsis of their bare souls, surrendering to love, yielding to the ecstasy of truth. This deep passion for mystery inspired my early years before fading, as childhood fantasies do, into the concerns of adulthood.

Four years ago, I stood at a crossroads in my life. I had walked the path of service since youth. I had traveled the world, worked as an executive, advocate, educator, and public speaker on issues of global poverty and environment, and pondered the customary questions about existence and my purpose on this planet. I had seen therapists, entrusted my body to the hands of healing artists, opened my soul in women's support groups, and cried in the silence of darkness in search of that unknown, unnameable force that an unfulfilled heart seeks with all its might.

I had learned to surrender and trust, and to complement linear thinking with the divinatory wisdom of the I Ching of ancient China. One day, I tossed three coins and opened a hexagram whose lines confirmed a possible path to a particular place at a particular time. I chose to follow that pathway. Within days, I met Roger Montgomery and, through him, became involved in an extraordinary system of self-knowledge based on the mathematics of the ancient Maya and Aztec civilizations. It is called Twenty Count. I soon found myself immersed in studying, living, and applying the principles of this system.

The visual design of Twenty Count embraces a chain of life concepts and processes, each represented by a number (one through twenty), positioned around a central zero, and plotted within three concentric circles that connote various dimensions of existence. Each number is an invisible world unto itself, a dynamic energy vortex that is continually shifting, transforming, evolving. This dynamism, however, is not haphazard; rather, every being can ascertain and travel a clearly distinct pathway corresponding to the optimum potential of Earth and universe in synchrony. Twenty Count as a system enables us to align our individual lives with the natural order. The result, which is also the process, can be measured in terms of the joy of fluidity, the deep security generating from decisions that embody the multidimensional alignment, and the peace of walking one's precise path on Earth.

As a woman who cares about justice, democracy, and peace, the very shape and form of Twenty Count resonates within me. The composition of the system as circles laid out in an orderly yet fluid tapestry is a refreshing departure from the predominantly vertical/hierarchical, or horizontal/flat organizations of many metaphysical and societal structures. Twenty Count is deeply inclusive. It acknowledges a place for everyone and everything. To me, this remains its central beauty.

Twenty Count connects us to all dimensions of life. It opens our imagination to perceive the world of spirits as an extension of ourselves, as the larger environment that nurtures and guides us. By gazing, projecting, and traveling through the unique window of Twenty Count to magical, spiritual dimensions, and by embodying the system's ever-expanding and demanding essence, I have learned to recapture the sense of wonder and the longing for the unknown that were induced in my childhood days by the poetry of Khayyam and Rumi.

Meditating on Twenty Count, for instance, my mind can easily rekindle the vision of a phenomenal time when, instead of depending on telecommunication and interactive computers, seers from distant lands compared notes in a silent communion of souls and rhythms. They didn't need verbal confirmation of one another's visions; they would simply intuit each other's spiritual discoveries and revelations.

I smile at this thought because, if someone had told me five years ago that, today, I'd be writing the foreword for a book about Twenty Count and musing about the spirit world, I would have dismissed that person as more than a little off. But now, this seems like a most natural act. Talking about and living with spirits has become part of my daily life's flow. This notion reminds me of Isabel Allende's book, *Paula*, in which for the first time this remarkable woman's life on the cusp of two worlds becomes crystal clear. There is no separation between the worlds; they relate to one another in a continuous dance of transcendence. Through this interaction, both dimensions co-evolve. I have found this principle to be the most enriching lesson and practice of my life.

As a professional in the field of organizational development, my work focuses on local and global institutions whose programs advance social, economic, and political change and promote human development. Sustainability is one critical issue for cutting-edge organizations. Through Twenty Count, I have discovered that properly tuned instinctual decisions possess a built-in consideration for sustainability. Every action taken at every moment with the instinct born of self-Earth-cosmos alignment carries within it the hologram of what Native Americans refer to as the "seventh generation." Spirit determines sustainability if humans are willing to hold that vision in place, steadfastly and with integrity. The mind needs not (nor could it possibly) do all the calculations about the impact of every human action on the future of the planet. The image of instinct in Twenty Count is that of a highly sophisticated computer that works with precision and protects itself against obsolescence by constantly expanding its own data base, a process that, in physical terms, can be likened to stretching the human envelope through continually embodying new spiritual and mental processes.

For Roger and me, partners in life and business, Twenty Count is an all-encompassing, nurturing environment for our daily existence. Twenty Count is a story of life, and Roger is, first and foremost, a writer. He is not a ritualist, a ceremonialist, a shaman, a therapist, or even a teacher in the usual sense of the term. In observing his process of writing this book, I was fascinated by his highly intuitive yet remarkably systematic mode of accessing, organizing, and communicating information. This approach appears to me as both the cause and effect of Roger's intellectual independence. He did not receive the teachings from someone else; he brought them to life through his own neurolinguistic faculties and personal discipline.

In working with Roger, I have observed that reverence is not an appropriate attitude toward Twenty Count. Actually, the adventurous imagination of a dazzled child seems the most effective means of entry into this illumined organization of the cosmos. I have also seen, again and again, that the moment I insist on interpreting or defining the Count through a linear, logical thought process, my interpretation or definition eludes me and evaporates. I end up circling around, asking and re-asking the same closed-ended questions. My inability to break through becomes frustrating. Twenty Count simply is not conducive to simple explanations. Therefore, instead of attempting further comment in this foreword, I offer the following poem as a welcome to *Twenty Count: Secret Mathematical System of the Aztec/Maya.*

Tao of Maya opens.
Aztec Jaguar roars
the gentle whisper of time:
Twenty; zero.
One breath, an eternity.
Some place in between,
wrapped in ecstasy of quest,
driven by rapture of love.
Or is it another force,
another word,
more potent than any other
that has ever come through
with that luscious smile of arrival?

XIV

Something new,
never seen before, nor felt,
yet as familiar as
bone marrow.
It's here. It's now. It is . . . .

Haleh Pourafzal
Berkeley, California
April 1995

Haleh Pourafzal is a transformation consultant and process facilitator with social change organizations. She has a master's degree in International Development and her work has included community empowerment in Africa, Asia, Latin America, and the Middle East. She is a contributing author to Non-Violence in A Violent World, *to be published in 1995 by California State University in observance of the 125th anniversary of the birth of Mahatma Gandhi.*

TWENTY COUNT

SECRET MATHEMATICAL SYSTEM OF THE AZTEC/MAYA

TWENTY COUNT

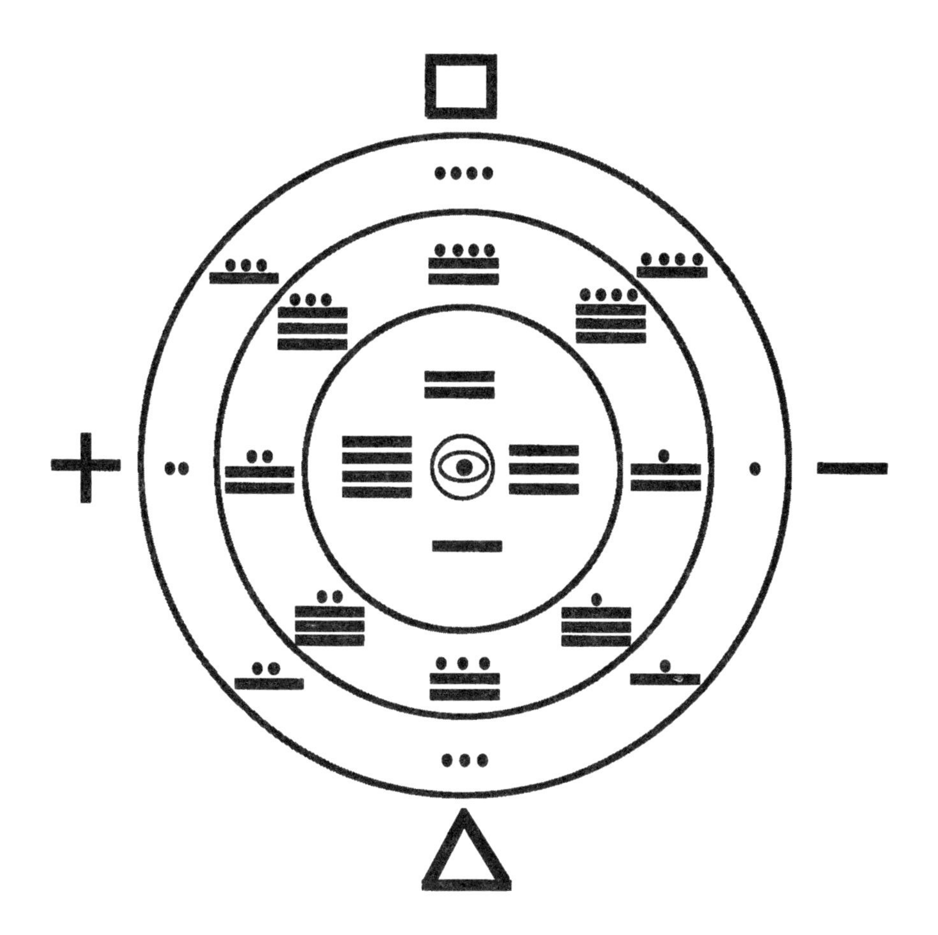

An ancient wisdom dwells within this thought atom. From time immemorial, we have sensed its presence and sought its guidance. In our current age, however, the atom lies obscured in remote recesses midst heart and mind. This writing traces a pathway of rediscovery.

Introduction

By ten years old, I had gleaned quite a collection of arrowheads from Indiana fields and pastures. Many sizes, many shapes, many types of stone. They went with me to college, then became my traveling companions when a journalism career beckoned me to the East Coast, and again when a spiritual quest led me to California. Today, these arrowheads keep me in touch with both my childhood and an ancient realm of wisdom.

These chiseled stone points are my friends, comforters, and teachers. During my quietest moments, they speak a silent language I understand. Their ancient presence has opened my inner ears to other voices as well, and to a powerful but little known creation story and inner guidance system known as Twenty Count, an empowerment tool as old as the arrowheads themselves.

Twenty Count is a model of self-discovery and growth for people in process, an accounting of the patterns of overlapping elements that inform us about who and what we are. The nature of its teachings reveals and informs through individual life processes, substantiating the view that a unique entry point and path of soul mythology exist for each person.

Years of meditation and direct experience have revealed the Count as an exciting source for opening life's potential. Every mental concept, emotional therapy, physical science, and spiritual discipline I've encountered through books, teachers, and travels fit like long-lost puzzle pieces within its parameters. As explained in the story to come, the production, timing, and guidance of this book are direct results of my own journey through the Count.

When studied in depth and applied with a sophisticated approach, Twenty Count works on a broader scale as well. Its organizational magnitude and operating principles are being used currently as both backdrop and method for effectively managing and accurately perceiving processes of change in a range of leading-edge organizations.

The roots of Twenty Count run deep into the ancient mathematics of Mesoamerica, especially the Maya and Aztec cultures that utilized similar sets of twenty pictographs roughly corresponding to tarot symbols or zodiac signs. On page 3, you will see a Twenty Count mandala that depicts a Maya Zero as the center; a layout of numbers one to twenty in Maya notation with a dot for one and a bar for five; and four basic universal mathematical symbols: minus, plus, triangle, and square. Structured around the relationships of logical thought and spiraling intuition concealed in this design, Twenty Count is a story of creation. It goes like this.

Through Zero, wind breathes life into four directions and four seasons, a creation dance of light and shadow. One is the fiery sun, the father spirit, the source of all we are. Two is earth, the mother spirit, fire's mirrored reflection cooled into firmament. Three is the emergence of plants as water flows through the earth. Four generates movement through outer awareness as animals evolve, seeing and hearing. Five is humanity, aware of itself, seeking identity.

Six displays humanity's journey from the ancestors' realm, our causes and effects in historic detail. Seven is the smoky world of dreaming, a panorama of raw phenomena. Eight is the tangling ties of pain and pleasure, our connections to matter, energy, space, and time. Nine offers the choice of whether we or the moons of history, dreams, and karma will rule our lives. Ten is the leap to higher awareness, our portal into the secrets of mind.

Eleven inspires us through the illumined vision of the stars. Twelve is the art of planetary organization, the awareness of self-based order. Thirteen is the transition into remembrance of direct perception. Fourteen awakens the power of instinctual wisdom of heart's logic. Fifteen reveals the holographic nature of existence.

Sixteen releases historical absolutism in favor of synchronicity. Seventeen disciplines the dream world with imagination. Eighteen activates creative power. Nineteen replaces humanity's burden of choice with inner knowing. Twenty returns to its source, the ancient heart of Zero.

From this simple creation story emerges a vision of our potential as human beings, regardless of whether we enter this life as ancient Aztec warriors or contemporary people. In fact, the very nature of our times makes Twenty Count especially relevant today. As information intake escalates constantly, we need more than ever a clear, cool source of grounded perspective to keep us both sane and productive.

Still, this perspective can be difficult to communicate. As simple as its story is, Twenty Count also can be complex and perplexing. For the past quarter century, I've been researching, contemplating, and meditating on the Count. A few seasons ago, I began this book. Friends would ask what I was writing about.

"Twenty Count," I'd say.

"What's that?" they'd ask.

"Well, it's a paradigm," I'd answer. "You know, a thought atom. A systematic collection of numerical ideograms radiating from the central vigesimal (twenty-based) mathematical scheme replicating continually throughout Mesoamerican calendar schematics. A self-organized map of the universe functioning as both context and method for connecting directly with the natural order of phenomena. A metaphoric system for gauging patterns of relationships in space and increasing efficiency in time."

"Yeah, uh, wow, that's something," my friends would say. "Want to get a pizza?"

Thus it occurred to me that I needed a better way to introduce Twenty Count to people who'd never heard of it. But the truth is that Twenty Count tends to both defy and transcend descriptions not based at least somewhat on personal experience. You need to get into it in order to understand what you're getting into.

It is possible, however, to take a different approach in drawing a bead on the Count. After a quarter century in its company, I can tell you what it's about, and that the process and the result are inseparable. Twenty Count is about the following:

- Practicing deep listening and direct perception in every moment.

- Taking personal responsibility for your own development. Finding comfort and opportunity in solitude. Opening your creative center and expanding your imagination to encompass the entire cosmos.

• Aligning outward and inward growth so that the inner being always takes on sufficient challenge to grow, but not more than it can absorb constructively as inner knowledge. Achieving fluidity through consistency of walk and talk.

• Connecting with the spirit world, awakening the animal power within, and recalling personal memory banks of your particular archetypal images. Trusting the unknown, as, for instance, when you hear the inner voice, yet discriminating against personally destructive fantasies.

• Pursuing transformation through precise, small, but impeccably accomplished acts. Being discerning and efficient while making instantaneous decisions and choices. Always giving priority to inner life and taking care of essential life details while eliminating futile actions.

• Allowing your own unique personal style to emerge. Basically being nonconformist. Freeing yourself from the pull of the world while being fully engaged in it. Carving out and shielding your sacred space without becoming an isolationist, and performing your daily responsibilities efficiently without being bogged down by the weight of distractions.

• Having courage to make choices that run counter to the established patterns if your instinct is strong and grounded enough to call you to action. This can include facing and overcoming fear of life-threatening diseases; fear of leaving one job before you have another waiting; fear of starting a new business during depressed economic times even though you have a great idea.

• Accepting death as a friend, not an enemy.

• Acknowledging and deciphering your unique path of service. Everyone has one.

• Living comfortably, with the beauty born of your own personal style, with freedom of flow in every moment, yet simply and with no extra baggage. Being ready at all times to complete your Earth journey without leaving behind either material or nonmaterial litter for others to clean up.

• Being productive in concert with others. Building partnership when things can't be done by one individual or group. Recognizing when your own capacity needs to merge with others', yet acknowledging your own field of existence—physically, spiritually, emotionally—and respecting the same in others.

• Honoring children and elders. Engaging in an alternative vision of a world where elders and children work hand-in-hand, elders raising and teaching children, and children serving and learning from elders.

• Greeting all beings, silently or aloud, with *In Lak'ech*, Yucatec Maya for "I am Yourself."

• Laughing.

Having related this list, I must point out that this book is not a step-by-step instruction manual for using Twenty Count or manifesting any of the above applications. Rather, this is a recounting of how I discovered its inner method through finding my own inner voice, while at the same time relying on that inner voice to guide the inner method.

The style of this book and the nature of Twenty Count itself is reflected in an introductory passage written by editor John Freeman for Jung's *Man and His Symbols* as a description of the spiraling form associated with Jungian argument:

> [The] arguments spiral upward over [the] subject like a bird circling a tree. At first, near the ground, it sees only a confusion of leaves and branch-

es. Gradually, as it circles higher and higher, the recurring aspects of the tree form a wholeness and relate to their surroundings. Some readers may find this "spiraling" method of argument obscure or even confusing for a few pages—but not, I think, for long. It is characteristic of [the] method, and very soon the reader will find it carrying him [or her] with it on a persuasive and profoundly absorbing journey.

Twenty Count and the Jungians travel in the same circles. But then, so does all the rest of life. Consider these thoughts on the language of the sky, taken from Dr. E. C. Krupp's *Beyond The Blue Horizon:*

> The sky speaks in celestial objects: the sun, the moon, the planets, and the stars are its vocabulary. The sky's grammar is what these objects do. They rise and set; come and go with the seasons; meet and separate; they circle or tumble or dawdle through heaven and time. The whole sky is a stage, and the things we see there are players. They make their entrances and exits, take their bows, and return for repeat performances.

Like Dr. Krupp's sky vocabulary and Jungian arguments, the concepts of Twenty Count come and go throughout the coming pages. With each repeat performance, new insights arise. The numbers and their values may be traced and pursued through the teaching wheels, the writings, and whatever associations you may make.

Also ahead, you will meet twenty spirits, guides who emerge from studying the Count. These spirits first appear as mentally dormant concepts, intellectually valid but lacking in inspirational power. Patience and perseverance convert these spirits into teachers for the spiritual quest. The Count thus serves as a storehouse for a synthesis of spiritual and intellectual processes, augmenting consciousness and focusing awareness.

The Count embraces the intellectual process of the heart, while also reflecting the demand for grounded, unblocked thinking. It is a way of perceiving and proceeding, incorporating elements of both ordinary logic and nonobvious insights, requiring the acceptance of hoary esoteric knowledge alongside the realization that neither the presence nor practice of such teachings constitutes an end in itself. It guides you into, through, and out of the maze of contemporary interpretations of ancient mystical disciplines. While acknowledging unlimited potential of mind, this process operates fully and effectively as a blueprint for intellectual strength and detachment from obsessions.

Guidance through the Count is available for both seasoned seekers and beginners alike. Your only prerequisite is to feel in your heart an urge to seek out answers about who you are and what you're doing here on this planet. Possibly you have spent decades traveling through therapies, ashrams, churches, workshops, and masters, continually learning but never locating your one special pathway. Perhaps you have read the Upanishads; the Tibetan Book of the Dead; the Egyptian Book of the Dead; Tao Te Ching; I Ching; Herman Hesse; D.T. Suzuki; Carlos Castaneda; Thoreau; Ramacharaka; Teilhard de Chardin; and Thomas Merton. Perhaps you have explored the psychedelic road of visionary expansion, or have bowed your head and prayed in synagogues, mosques, temples, churches, sweat lodges, deserts, canyons, and rainforests. If so, you can count on being challenged by the Count to create a crystal-clear alignment of your experiences in order to enhance clarity of thought and focus. But be forewarned: prolonged exposure to Twenty Count has been known to generate unprecedented creativity and effectiveness. You may be required to put all your study and learning to use.

On the other hand, if you've never meditated one moment, never studied religion or philosophy or science, never set foot upon a single esoteric path, you may be at great advantage as you enter the Twenty Count experience. Your mind may be open and free, unprejudiced and nonjudgmental, and the simplicity of these circles, numbers, and ideas will make great sense to you. An aboriginal calling may sound deep within the soul. I have seen two people on separate occasions rise from their first meeting with these spiraling wheels and start painting extraordinary mandalas representing their own unique visions of the Count.

This is not to claim, by any means, that everyone who encounters Twenty Count is transformed. I've also seen people look, listen, shrug, and walk away. Expressions of indifference often fascinate me. I recall that same shrug from an angry man at a Maya Angelou poetry reading; from an unimpressed museum visitor in a roomful of original engravings by William Blake; from a tired woman listening to the "Hymn to the Sun" from *Le Coq d' or,* Rimsky-Korsakov's mystical Russian opera.

But I refused on those occasions to believe these people were untouched in the depths of their being by some minute shiver brought on by the raw beauty of the poem, or the wild grandeur of the vision, or the aching melody of the music. If the shiver did not surface consciously, then it ran somewhere in the deep, dark, shadowy unconscious. In the same light, I do not believe that anyone who has encountered Twenty Count ever walks away without some slight shift in perception and inspiration. Perhaps long after glancing at the spiraling circles, even the shruggers will dream an elusive image of radiating numbers and spirits while a silent, patient voice whispers. And perhaps that voice will return at the precise moment they most need guidance on life matters.

This inquiry into Twenty Count explores its seemingly endless river of information and its methods of converting that information first into knowledge and then into wisdom. Guiding spirits dwell within the Count, and every reader is likely to discover radically different aspects of both spirits and pathway for every number. That discovery is imminent now that you have entered the path.

Twenty Count reveals challenging and mind-expanding guidance for comprehending life's patterns. Using the Count involves reading, study, and contemplation in order to confront confusion until it transforms into clarity. Commitment is required, but once mastered as a tool for creative productivity, the Count offers unlimited applications. By accessing its guidance, you can achieve an expansive view of existence and an unprecedented approach to contemporary life choices. I now invite you to travel with me in the coming pages on a path of inquiry within your own mind, listening to your inner languages, and visualizing individualistic patterns of revelation through inward contemplation and dreaming.

Read on, quest well. Enjoy your journey as a search for clarity, a journey into identity, and a return to the ancient heart.

SERPENT AND ZERO

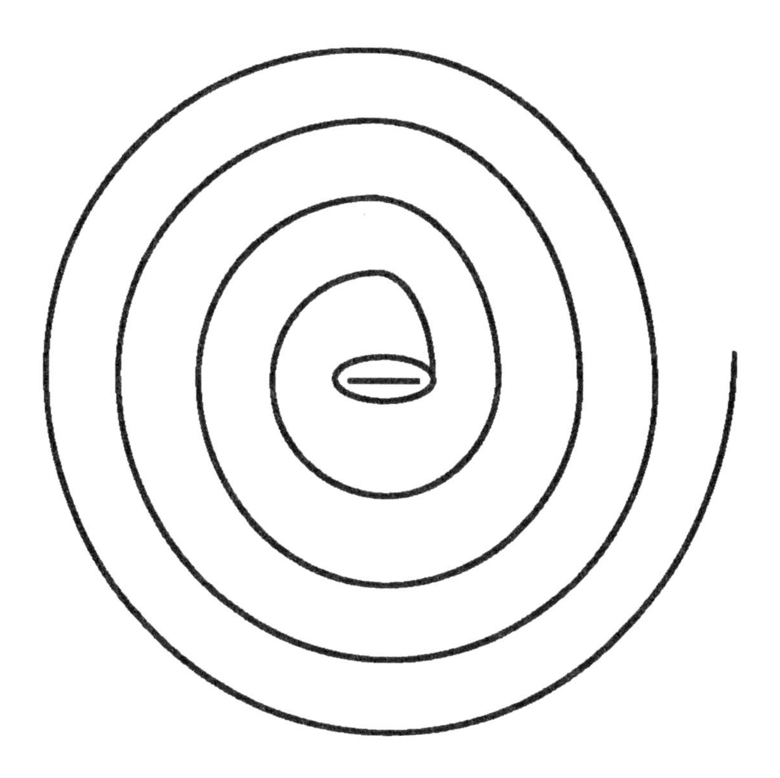

A motive
for murder
for men
without women.

For the Scribes

Long ago on a dark day, invading priests collected the beaten-bark library of ancient scribes and burned the books to cinders.

These pious conquerors were horrified that the native culture honored the serpent–the coiled spiral–as a spiritual symbol. As children, the invaders were told a serpent corrupted them through a woman. They believed it literally, the great metaphor dangling powerless amidst their obsessive passion for honoring a power greater than themselves. As grown men, they rejected women and cloaked their bodies to search for redemption by converting all creation to their solitary vision of misery and lost virtue.

Compounding the native spiritual and intellectual transgressions was yet another symbol, a seashell-like image derived from ancient perceptions of the Milky Way–an oval with a line through it. The symbol surfaced everywhere, repeating continually throughout mathematical accounts that soared back 400 million years. To the scribes, the symbol represented Zero. To the invading friars, it looked like a vagina.

Entranced by self-righteous zealotry and fear, the priests condemned a culture because they could not see its vision. They piled up and ignited an eon of mathematics, theology, art, astronomy, history, literature, mythology, architecture, and navigation. To expunge the spiral from native souls, the invaders established sacred policy of killing, converting, torturing, mutilating, and enslaving. The murderous violence, a force more enduring than the invaders themselves, continues today.

On that dark day of burning, flames shrieked red and orange and high, cursing into oblivion pages laden with observations and perceptions of learned scribes. Black smoke billowed and curled into clouds. Yet even as markings scorched away and bark paper crackled to black ash, the essence of that recorded knowledge also rose to the sky. Winds gathered that essence, reshaped it, and bore it aloft for centuries.

Now a time has come when many people no longer fear the serpent. The small mind of fervid fantasy has commenced its inevitable lapse into fetid decay and has lost its stranglehold on the creative soul of global culture. The winds breathe the lost wisdom back into our midst, and scholars and seekers have reawakened to long-silent secrets. Like the old ones, we study the past to prepare for the future. We see silent patterns persisting throughout the web of our universe, filling us with awe.

This writing, to whatever degree and in whatever dimension it reflects the spirit of that ancient wisdom, is dedicated to every scribe whose work went up in smoke. You are not forgotten.

May we never again fear the serpent.

May we open heart and mind to the journey from and to Zero.

May the killing cease.

AZTEC SUNSTONE

The Source

Watch, my friend, and see a wondrous sight, a spinning rainbow of twenty circles of light bursting out of the fleeting gleam between elder and child, from passionate lovers one to another, from Earth to her myriad creatures. Revelations of soul wisdom dance and sing in the blinding primordial flash between birth and death. So long as you breathe on this planet, you will quest for an ancient truth moment after moment, even though I tell you that truth this very day. You see, you will not remember these words, and therefore you must always seek to hear:

"The source of Twenty Count is the singular light of soul."

Mysterious inner voices speak to us. Just before we dream, or at the foot of waterfalls, or when we stand naked on midnight mountain tops, messages flow into our minds from deep within our hearts. If we listen clearly and discriminately, we are comforted in spirit and guided effectively through the physical world. But we must stay alert or fantasy can deceive and delude us. Life has taught me this simple truth.

Precisely when or even how I first heard those words about Twenty Count and the singular light of soul is still not clear to me. But one day, maybe twenty years ago, my mind was buzzing with: "The source of Twenty Count is the singular light of soul. . . . The source of Twenty Count is the singular light of soul." My feeling, on reflection, is that a desperate but sincere plea for clarity during years of meditation awakened a playful ancient essence who decided to adopt my orphaned inner life. The entire message inspired me. I wrote down as much as I could recall several times, but I kept misplacing my notes. No one I asked had ever heard of anything called Twenty Count. Nothing had been written about it. In time, even the last line faded from my memory.

I had no idea how difficult it would be to seek out the source of that message. The search evolved into a journey of years and roads, teachers and friends, solitude and companions, questions and answers. Writing about this quest provides me the opportunity to build a framework of personal reference points for the many and varied teachings encountered along the way. Though this book takes an autobiographical form, its content is dedicated primarily to the story of Twenty Count.

To fill you in about both the search and the Count, I must take you back in time to provide perspective. Back far beyond my first inner voice, back hundreds of years to the high civilizations of the Aztec and Maya peoples of Mesoamerica–a time frame that doesn't really feel that long ago to me anymore, perhaps because I'm sitting here with one eye on the present and the other on antiquity.

Marvelous mathematical icons from two dramatically diverse civilizations stand next to one another on my desk. One is a sleek, purring computer on which I type. The other is a foot-high clay reproduction of the ancient Aztec Sunstone, an image that riveted itself into my imagination from a glance at a textbook reproduction shortly after my Twenty Count message.

Computer and Sunstone are a study in contrasts. The computer arrived here two years ago as a state-of-the-art extension of the applications of logic, a distant cousin of the aerospace computers that guide our spaceships. Already, however, our era of exponentially accelerating technology has outmoded this model. By the time you read these lines, it may well be obsolete, its inner capacities no longer compatible with the demands of the times.

The original Aztec Sunstone—twelve feet in diameter, four feet thick, and weighing twenty-four tons—was discovered in 1790 near the site of El Templo Mayor in Mexico City, and has been dated back to the fifteenth century. Still, the Sunstone's spiraling display of twenty pictographs called *tonalli,* similar to zodiac signs or tarot symbols, remains as relevant today as in the era of its carving or on the day of its unearthing. Obsolescence does not threaten the spiral because its data base interfaces synchronously with the human mind and heart. Technology, despite its awesome accomplishments, is powerless to duplicate this prime attribute of enduring relevance.

The Sunstone is considered a highly complex hieroglyph of Aztec cosmology that claims we're living in the fifth age of Earth. Take a look at the rendering of the Sunstone on page 19, and you will see the central face of the god *Tonatiuh,* the primordial creative deity. Radiating outward are images of the four suns of previous ages. The next ring, a graphic representation of Twenty Count, contains the *tonalli.* The outer circles depict various celestial symbols embraced by two fire serpents.

Roots of the Sunstone design stretch back through Mexican antiquity to the Classical Maya and their wondrous mathematics that utilized a dot for one, a bar for five, and a shell-shape for zero. Operating from a numerical base of twenty rather than our current base of ten, the Maya achieved incredibly sophisticated calculations reflecting a breathtaking overview of time. The Maya calendar known as the *tzolkin* revolved around a twenty-number cog wheel, as did the later Sunstone. Other Mesoamerican cultures leaving at least remnants of similar calendars range from the Olmec, the mother culture of Mesoamerica, to the Teotihuacan and Toltec.

Today, the hearts and minds of contemporary seekers around the world are pulled powerfully toward the ancient Aztec and Maya. Images of and references to these ancient cultures are surfacing in the inner eyes and voices of an increasing number of people, possibly because our contemporary collective unconscious has imbued our memories of these cultures with mystery and romance. Or perhaps it is because the same inner voices that once spoke to these people's minds from their own hearts now actually speak to our minds.

My own affinity centers on the mathematical system. The first time I saw the Sunstone photograph and related it to Twenty Count, it seemed merely a complex variation of the Native American medicine wheel. Then, through my years of meditation and application, the Count's own inner teachings emerged as the very basis for the wheel. Eventually, I learned to appreciate the Count as a powerful but elusive process for grounding into everyday reality an extraordinary scope of universal teachings.

The medicine wheel paradigm itself came into public awareness during the 1960s when various tribal spiritual leaders opened teachings from their own lineages to an enthusiastic but sometimes unfocused generation of seekers. *Black Elk Speaks*, a book of Oglala Lakota revelations, surfaced as a cult classic after being ignored for decades. As years passed, seekers began to recognize teachers such as Rolling Thunder, Sun Bear, Twyla Nitsch, Mad Bear, Evelyn Eaton, Semu Huaute, OhShinnah, Bear Heart, Dhyani Ywahoo, Hyemeyohsts Storm, and Brook Medicine Eagle. These elders and teachers spoke of stone circles created and bequeathed to us by ancient peoples throughout the world as visual representations of the medicine wheel.

The medicine wheel's cross-within-circle image encapsulates philosophy, psychology, spirituality, numerology, and cosmology. The four quadrants of the wheel describe cycles of all we know. They represent east, south, west, north; spring, summer, autumn, winter; morning, mid-day, evening, night; birth, youth, adulthood, death. The wheel teaches, guides, heals, and reveals. Nothing simpler or more complex exists.

Life's spinning patterns stamped their imprints in the minds of artists of long ago. Centuries before Carl Jung visualized and postulated archetypes; before James Joyce twisted and reshaped the English language into a plaything opening to the other side of writing; before Joseph Campbell synthesized parallel symbolism from throughout the world into a unified flow of mythological perception–Maya scholars created their own geometrically ingenious designs and inscribed archetypal day signs in codices and on stelae. They meticulously recorded their mechanics of wisdom, the flow of natural relationships, the process of growth, and the semiotics recalled from the primordial flash. They computed the year as 365 days. They coalesced philosophy, science, mysticism, and art. Chronicled within their system of relationships are patterns of matter, energy, space, and time that define our cosmos. These patterns reveal blueprints of our humanity: rationality, physicality, emotionality, and spirituality.

During my own travels in Mesoamerica and through poring over writings about Maya calendrics, I have been struck by one lyrical poem that stands head and shoulders above all analytical efforts at exploring this calendar system. In 1957, Mexican Nobel laureate Octavio Paz penned a 22-page, single-sentence poem called "Piedra de Sol"–The Sunstone. This mind-blowing interweaving of history, visionary dreams, erotic love, time, and memory created a brilliantly lit stage on which future interpretations of the paradigm could assume their unique roles. I also believe now that it was in an introduction to the English translation of the Paz masterpiece during my college years that I first encountered the written term "Twenty Count."

Understanding Twenty Count as the key to the Aztec/Maya system of self-knowledge, however, required Paz to envision numbers in the same light as did these ancients. For them, numbers represented everything in the universe, both that which could be seen and measured, and also that which could not be described in the world of form and limitation. Mathematics served as their extreme reduction of philosophical thought, a system of self-knowledge with universal application, based on long-term scientific observation and expressed through religious feeling and artistic creativity. The Maya were the first culture to use the concept of zero; that is, allowing something to stand for nothing in their notation system. It's not an obvious or easy starting point.

The first widely publicized attempt to examine the Mesoamerican concept of twenty numbers as a paradigm occurred in the middle years of this century and involved English explorers,

American archaeologists, and some imaginative journalists, all obsessed with a controversial, life-sized skull carved from quartz crystal. This skull reportedly was discovered in the Maya city of Labaanatum in British Honduras (now Belize) in 1927 at the height of archaeology's popularity, when scientists were desperately seeking publicity to generate funding for trips into jungles and to pyramids worldwide.

Since the workmanship on the skull was unlike any other known Maya artifact and since quartz effectively resists carbon dating, it is impossible to verify the skull's origin or purpose. Rampant speculation about its origin has ranged from the Knights Templars of Europe to ancient Japanese seafaring monks to the Maya's own distant ancestors. One enduring theory claims the skull not only preceded the Maya but also was the focal point of a twenty-number divination system that evolved into the Mesoamerican calendars. This theory has spawned a number of newspaper and magazine articles over the years.

Though researchers continue to explore extensively the depths and secrets of these ancient peoples' shamanistic beliefs, the significance of Twenty Count is seldom scrutinized. Analysis of the twenty Aztec and Maya day signs and their possible meanings remains more or less confined to books on Mesoamerican astrology. When other writers refer to the Count, they do so only in passing, sometimes offering brief commentaries on its structure and interpretations.

British author Kenneth Meadows displays an illustration of the count in *Shamanic Experience*, and explains, "It is not possible to provide an interpretive analysis [of Twenty Count] because such insight and knowledge is personal and comes shamanically." Meadows studied with shamanic teacher Harley SwiftDeer

Reagan, who has developed an extensive system of Twenty Count-based teachings offered through the Deer Tribe Metis Medicine Society of Scottsdale, Arizona. In *Star Warrior: The Story of Swift-Deer* by Bill Wahlberg, SwiftDeer describes the Count as one of the "primary tools to facilitate ways of organizing freedom. From the teachings of the Twenty Count, one learns that everything in the universe has order."

One feature of the Count is that each of the twenty wheels contains twenty sub-wheels. Mastering a concept opens that principle's wheel so that, like the universe itself, the paradigm expands. In *Fools Crow,* biographer Thomas E. Mails reports that Frank Fools Crow, the acclaimed Teton Sioux holy man, attributed his miraculous healing powers to a group of 405 spirits. To students of Twenty Count, the number 405 represents a spiral of twenty wheels, each containing twenty sub-wheels, plus five basic wheels for the four directions and center (20 x 20 + 5 = 405).

Aside from Twenty Count's origins and its resistance to analysis, this simple diagram with its basic philosophy of aligning history, dreams, relationships, and choices has influenced my thinking far more than all the teachers I've encountered and all the books I've read during my half-century of life. Had I been born in Tibet, I might have found inspiration in a tanka; in Australia, in a didjeridu. But I come from Indiana, land of the ancient Adena and Hopewell cultures, where stone arrowheads surfaced in the soil and Twenty Count took form in my awareness.

Throughout this landscape of my youth, the most inspiring object for my imagination was Ohio's Great Serpent Mound, just a day's journey away. The world's largest effigy mound, the serpent dates from the first century B.C. and writhes more than 1300 feet, at a height of about three feet, atop a prominent hill. Many smaller mounds dot the area as well. Artifact evidence of these Mound Builders of prehistory has led archaeologists generally to conclude that they stemmed from the Olmecs. But, as Frank Waters speculates in *Mexico Mystique,* since the Adena people can be traced back to 2450 B.C., centuries before the earliest known Olmec date, it may be that the Olmecs actually descended from the Adena.

In either case, the artistic earth-formation essence of the Mound Builders bespeaks the same primordial source as the much later, mathematically precise pyramids of the Maya and Aztecs. It's my guess that this deep Earth-based feeling is what first connected me to these ancient formations and their applications to the present and future.

So far I've related precious little about Twenty Count, about the nature of its actual teachings, and about how pursuit of the source of my mental ramblings became my life's focus. That story is a spiral tale to be spun in the pages ahead. I can tell you now, though, that whenever I sit and contemplate the face of *Tonatiuh* on my miniature clay sunstone, I see the Count and I hear the Count and I feel the Count. I've come to appreciate its essence as my personal guidance system in the ever-present search for awareness, and I treasure its teachings as evidence of the ancients' mastery of enduring relevance.

But I still can't remember where I heard that message about the singular light of soul.

MYSTERY WHEEL

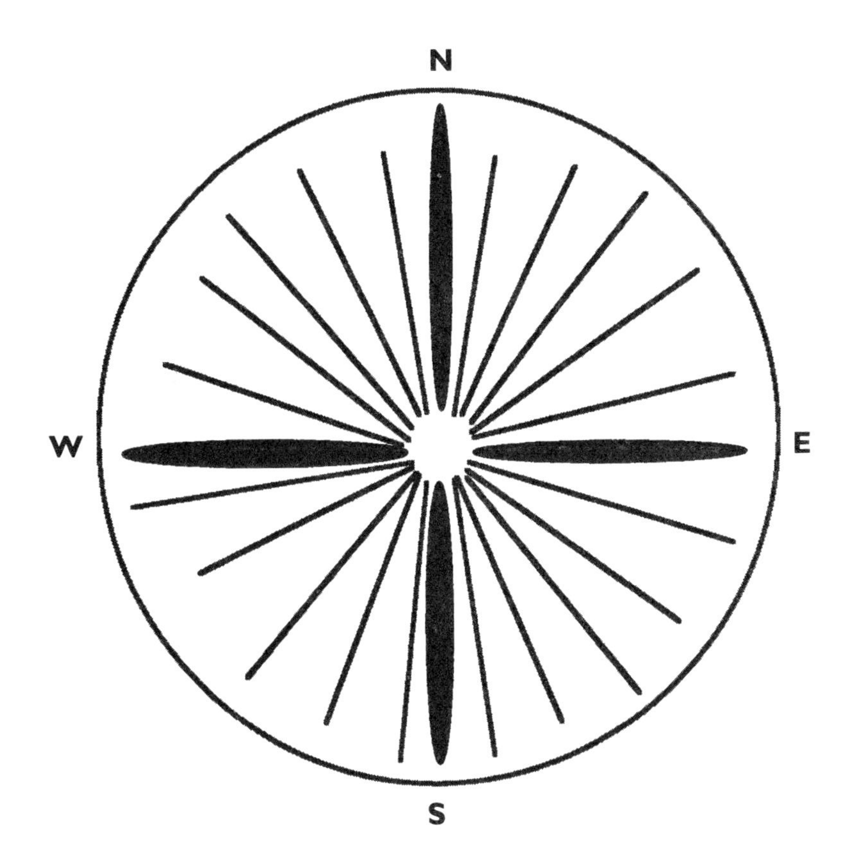

Two Sides

One Center

Four Directions

Twenty Rays

Algonquin Angst

Whatever occupied past years of my life, I let it go. Failure and success, identity and desires, I give up. Nothing fanatical, but simply no other way. I surrender. After all the study, practice, reading, and thinking, the alternating solitude and ineptitude, exhilaration starts from scratch. I'm dying, yet vibrantly aware.

Existence, as we dream it, floats in a fog of instantaneous forgetting until we recover the secret of spontaneous remembering—our primary purpose. One then returns that remembering to essence in exchange for the opportunity to summon freedom. Regulating this intake of memory and the ensuing swap for freedom presents the constant challenge, always directly ahead and attainable solely by manipulating perception. When freedom comes, we must be ready. To achieve and integrate this staggering barter of loss and gain becomes my work, my life.

In late winter 1983, I squatted on the barren precipice of Algonquin Peak and committed to this release. Midday winds whipped my face and swirled purple clouds into beasts of eons past. Nothing human-made appeared within the sweep of wilderness and low mountain. The steep, two-hour climb through birch and pine elicited a chilled clarity atop the gray summit. I had thought to hike on to Mount Marcy, but this old plateau captured me.

My temples throbbed with the ache that had impelled me onto this road through the timeless Adirondacks. That ache, the need to meld my daily world into a more relevant spiritual landscape, tormented me. Despite my most focused intent, the necessary steps loomed ahead as indistinct as the elusive creatures above.

I'd yearned for years for guidance into the fleeting tunnel of wonder called Twenty Count, a structure I acknowledged as embodying the pursuit of self-knowledge. Scanning for reference points, I discovered too many. Merely opening a book or commencing conversation summoned up hosts of possibilities. Every friend recommended a therapy. Philosophers, poets, healers, and teachers offered many avenues for pursuing reality and unraveling its essence. But this climb in the wilderness quieted me and revealed a purer, simpler vision of the endless paths and crossings. I knew deep-down that I needed nothing more than the bag of stones in my backpack.

The throbbing eased in my temples as I opened the leather bag of arrowheads gathered in my youth from Indiana fields. These sharpened stones of long-forgotten artisans preserved their creators' ingenuity in varied sizes and shapes. Occasionally I gave an arrowhead to someone special. One friend, in return, presented me the leather bag, both sides painted with rough medicine wheels of twenty-four irregular rays emanating from center to outer rim. That night, I counted forty-eight arrowheads, and the bag became my portable altar. Its mystery wheel image centered my meditation focus, grounding the deepening relationship of my self and my search.

Spreading a circle identical to those on the worn bag, I spaced twenty-four stone points. At the center, a tiny carved crystal skull. Four arrowheads marked the directions of the traditional medicine wheel, and another twenty created the circle. What each stone represented I didn't know. Enchanted amid rising winds and snow clouds over Algonquin's tabled surface, I was oblivious to the subtleties of the mystery wheel. Still, my limited perspective inspired a productive approach.

If a spirit dwells in each arrowhead, I reasoned, then a committee of spirits could guide me. If all merge, one powerful force forms. That could answer my need. Repeating my previous vow of release, I affirmed the refusal to cling to any element of the past, no matter whether it seemed positive or negative. My frozen neck relaxed. Slow breathing into the landscape softened the sharp wind and flashed a reflection of the path up to and beyond my moments here on Algonquin.

Years of spiritual adolescence had elapsed since an early introduction to fasting, sweat lodge, and vision quest. Next had come the cultivation of focus to generate inner awareness and change. During this period, youthful motivations of anger and fear lost their charge, and an entrenched self-pity simply self-destructed. Feeling disciplined but pointless, I mastered techniques and devotions and legends, playing roles of both adult knower and child learner. Unknown realms emerged rich in opportunity, challenge, and confusion. Elements of progress overlapped recurring relapses and encompassed years.

My search, as for many of my generation, had heightened during the 1960s' psychedelic drive toward consciousness. Blending disciplines from global cultures produced a sense of world citizenry whose allegiances were due humankind and the planet rather than nations or religions. We studied at universities, protested established values and a deadly war, and sought insights into life's magical secrets. We traveled with backpacks to San Francisco, Cambridge, and Katmandu. We relaxed our dress and grooming. A new music emerged as the pulse of the era. When that music stopped, many found a return to normality difficult if not impossible. But seeking paid a very low wage.

For fifteen years I balanced journalism with a search for clarity. They did not blend well. For every rational step my material life advanced, my perceptual capacity for the big picture receded. The professional career and internal quest coexisted and intertwined until fading into the indistinguishable. Reporting and writing grew agonizingly boring and unrewarding. The profound and undeniable inner need transformed into unyielding and frustrating self-reflection. That which interwove career and search one day demanded choice. I quit journalism.

A way of knowledge sometimes arises after a substantial trek through a demanding discipline. Knowledge as a path relies on honing one's perception to capture and comprehend the illumined insights of the world's light and shadow show. The mind resolves its dual components into unity and integrates experience, abstractions, symbolic patterns, and metaphors in order to apply and pass on the lessons learned. When fully present, the way of knowledge reveals years of preparation as merely a base of applied time and concentration. This perspective empowers the application of life teachings to confront the mental barrier.

That mental barrier had become my companion by the time I left journalism. Inspired by mythologist Joseph Campbell's courage to retire to rural Woodstock and read for five years while the country panicked for pennies at the onset of the Great Depression, I set aside two years during prosperous times near the Rhode Island shoreline for reading, yoga, and martial arts.

The early months of contemplation brought into focus a vague, haunting face as backdrop for my dream world, a craggy visage of weathered skin and sparkling eyes that projected strange but familiar geometrical constructs through my dreams. Many of these dreams related directly to my mystery wheel. Though depths of meanings bypassed me, I recorded what I could. The dream presence guided my reading and disciplines, and also forced me to acknowledge an emerging side of my nature that could be vividly described though barely discerned.

Then, suddenly, the dream face surfaced into reality. One day outside Amara's Restaurant amid the small shops on funky Wickenden Street in Providence, a burly man quick-stepped his way through snarled traffic, an agile dancer whose grace belied his linebacker's build. He wore his black hair close-cropped, a white-beaded headband with a blue mandala centered low on his brow. His skin was deeply burnished brown. Someone called to him, and he smiled a heart-smile and shouted a greeting. From a distance, I guessed him thirty-five to forty years old.

Red Thunder Cloud was an herbalist and healer from the Catawba Nation of South Carolina. Amara's was one of the many stops on his itinerary of healing up and down the Atlantic coast. He showed up every few weeks to see patients, deliver herbal preparations, and tell stories about everything from fist fights of his youth to how chipmunks got their stripes by scratching each other's backs.

Meeting Red Thunder Cloud opened me to the power of his grounded self-confidence, enthusiasm, and humor. "You know how old I am?" he laughed without my asking. "I'm sixty-two. But I still play basketball with the kids." He emphasized meanings with easy gestures of thick, powerful hands adorned with a family of time-tarnished silver and turquoise rings. The sleep-induced image paled in contrast to the man, and it vanished forever from my dreams.

The Catawba healer laughed easily and often, and loved to tell how his grandfather in South Carolina named him for a spectacular cloud formation at the moment of his birth. The last living speaker of the Catawba tongue, Red Thunder Cloud recorded that language through stories and songs of his tribe in a series of tapes for the Smithsonian Institute. Today, these tales and chants evoke in the listener a haunting remembrance of long-ago campfires and a bright-eyed child listening and watching as elders recited legends and led rituals for a tribal people faced with encroaching obsolescence.

A man of enormous energy, Red Thunder Cloud drove his van–customized with Native American symbols–along the Eastern seacoast, carrying herbs and healing to Native American communities and anyone else in need. He was a master storyteller, drummer, and singer. Teaching came easily to him, be it dancing, wilderness living, cooking, or the wisdom ways of herbs. He sold herbal teas and, when he had a supply from South Carolina, jugs of delicious Catawba honey. He crafted pipes, performed ceremonies, and actively supported the growing Native American rights movement.

Lecturing, he opened talks on herbal healing by welcoming: "Ladies and gentlemen–and members of the American Medical Association." AMA skepticism, however, did not bother him. As a healer, he'd built an impeccable reputation. Many doctors attested to the healing power of his herbal blends. Healing arthritis was a specialty, and he offered several different formulas.

I thought that interviewing mystical teachers had readied me to meet Red Thunder Cloud, but his impact proved different than anticipated. He became a friend, not a guru. His teaching flowed to the world rather than to isolated individuals. We spoke often about many things, but his wisdom offered itself in abstractions and subtleties, not instructions. His presence did, however, encourage me to embody his effusive joy. Red Thunder Cloud demonstrated that one could travel a straight line through life. Amid a society of ambitious, unclear people struggling with endless ego concerns, he strolled easily ahead as though no question existed about his next act. This dancing, singing, laughing man showed himself to the world with total protection but no defenses. Because he held nothing back, life opened for him.

By the spring of Algonquin Peak, I had accepted Red Thunder Cloud as a buoyant, inspirational imparter of an otherwise silent knowing. Today, we live a continent apart and we've not spoken for a long time. Neither of us likes telephones. But my inherited awareness of his spirit has established a daily process of comprehension and renewal.

The Providence experience prepared me for the decade to come. Embracing years of experience and teachings, I hungered for a singular vision. Pausing on this Adirondacks site en route to a new life in California, I pondered a return to the East. Red Thunder Cloud had entrusted me with three bags of sassafras bark to deliver as gifts to Native American healers on the West Coast. He said he'd see me again. But he couldn't say when. "That," he laughed, "is the Indian way."

Meanwhile, my need to establish a path of clarity had grown to near obsession. In response, a mental flow sprang up through my combined dream and waking states, and stewed in my mind. Slowly, more and more precise guidance emerged to focus my attention on a sustainable worldview. Bit by bit, I had become more aware of my own inner resources.

Two hours of reflection lapsed on Algonquin Peak. Like travelers and dreamers from forgotten centuries, I'd tripped into a pristine vista, opening easily, painlessly. Within that opening a force settled, a strangely discomforting yet uplifting convergence. For no discernible reason, I laughed aloud. No living creature answered, and I gathered the arrowheads into my pack. From a silent source sprang the urge to descend. The high and unbounded loneliness of Algonquin spurred me into a jog away from the further path to Mount Marcy and downward to Lake Placid and my rented cabin.

That night, I built a fire and watched a spring blizzard outside my window. Sleep and dream crept in amid tiredness and warmth. Mind projected an initial image of the inner workings of a computerish complex with rows of files, gear wheels, and flashing lights, somehow both scientific and toylike. An impersonal and mechanized device of extraordinary efficiency. Deliberately, the machine demonstrated its functioning logic, its self-perpetuating rationality. Then the static image slowly reassembled itself into a bellows of breathing. Feelings surfaced, fangs grew, ears pointed up, sinews stretched, and silver-black fur shimmered into essence. Violet eyes glowed.

An ancient Adirondacks roamer gleamed across centuries to curl up at my fire, ease into my dream, and offer a guiding intelligence for the journey ahead. Wisdom also stopped by briefly before vanishing, only to reappear and vanish anew in seasons to come. Fortunately, my ancestral friend remained constant, a trusted companion whose guidance transformed the vast maze of our world into a comprehensible progression toward a nurturing whole. The mazeway–twisting, narrow, tricky, and evasive–leads to a landscape where mystery and reason meet, where sustainable awareness is the only price for freedom.

The voice sounded clearly:

"Relax when you sleep. Learn to be simple. Accept my presence as logic. Wisdom is awareness with knowledge, the gift logic offers. You must learn the art of remembering that you have forgotten. Let me teach you. Do this: Perceive my message, follow my lead, watch your self appear to act, recognize your developing attention, enjoy it, and return to your perceiving. Beyond that, forget all that went before. In this way, the barrier will dissolve. The mystery will unravel. Now listen closely, for I must tell you my name."

Moments passed. I drifted into deeper sleep.

"Wolf. I am Wolf."

VISION OF THE ANCIENT HEART

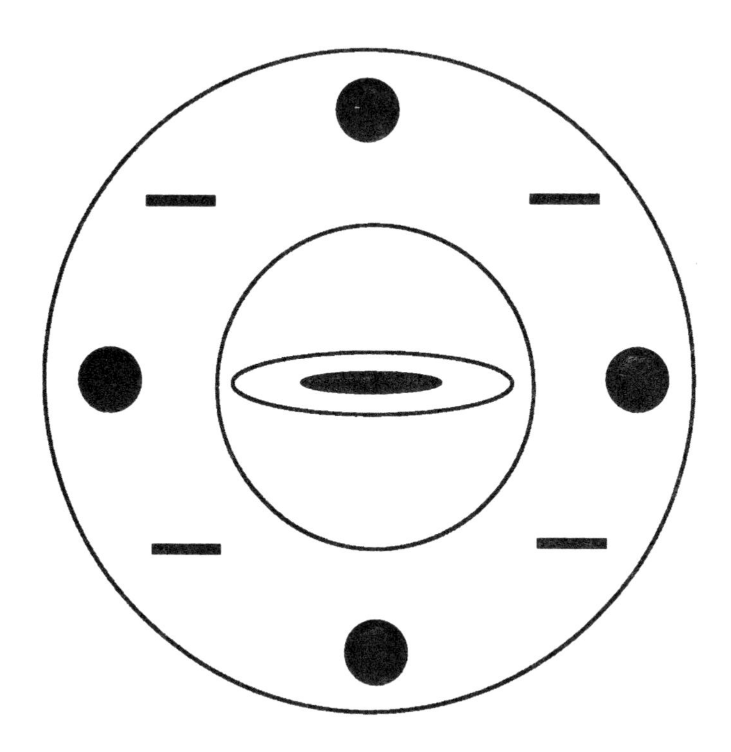

Vibration sings us to life.
Alignment stabilizes us.
Wind activates us.
Order gives us purpose.

Long Count to Twenty

Modern [humanity] likes to pretend [its] thinking is wide awake. But this wide-awake thinking has led us into the mazes of a nightmare in which the torture chambers are endlessly repeated in the mirrors of reason. When we emerge, perhaps we will realize that we have been dreaming with our eyes open, and that the dreams of reason are intolerable. And then, perhaps, we will begin to dream once more with our eyes closed.

– Octavio Paz, *The Labyrinth of Solitude*

Under Wolf's supervision, I watched Twenty Count emerge. It outlined studies, guided processes, and refined perceptions. Dimensions of the original mystery wheel increased in scope and transformed into the vision of a cycle of teachings known as the aspects of Zero.

"Twenty spirits live in the Count," Wolf said. "Each spirit has its own home, its own purpose, and its own idiosyncrasies. To function effectively in the world of these spirits, you must first grasp and merge with the aspects of Zero. But I warn you now that this merging will result in the simultaneous birth of twin realms of reality and fantasy, and you must be aware of this duality every step of your path. You must stay alert and play with duality for a long time before you can understand its full power.

"There's a lot to learn about Zero and duality," Wolf stressed. "Zero is an appropriate starting point for questing into Twenty Count. The journey is simple, yet seldom apparent. It is safe, but can threaten your deepest core. It's a process of ultimate change in order to achieve lasting stability. The road passes through the reality realm and smoke knowledge. It is the road to completion, but it's a long count to Twenty."

Wolf delivered the teachings in a dry, almost pedantic manner, sounding like a textbook on tape rather than a wild intellect turned loose to howl. He advised me to stay focused on the explanations because in time they all would make sense.

"It will also get livelier," Wolf promised. "For now, I'm studious and philosophical. That's to keep you from getting the idea I'm just some wandering werewolf from an old horror flick. Go ahead. Ask me some studious, philosophical questions."

"All right," I said. "What's Twenty Count?"

"It's a metaphor for the study of consciousness."

"A metaphor? Why a metaphor? Why not speak directly?"

"Because direct perception is overwhelming for you at this point," Wolf explained. "You need to work on yourself first. That's the purpose of Twenty Count. To recapture your vision of the ancient heart. To remember that vision as the start of your quest. You also will finish there. Think of your mystery wheel of arrowheads with its center point and twenty-four rays. Go ahead, close your eyes and get the picture.

"Now focus in on the center. Watch it become the Maya Zero, the sideways oval of the Milky Way. And at the very heart of the oval itself, see a smaller oval of black. That's the unknown source. I call it the ancient heart. Though you can't tell it at this instant, four basic elements make up that vision: vibration, alignment, wind, and order."

Envisioning the mystery wheel, I could see the four elements form in a manner consistent with Wolf's descriptions. As he spoke, the vision shimmered into pulsating action.

"Think of it this way. Vibration sings us to life. Alignment stabilizes us. Wind activates us. Order gives us purpose. I'll explain.

"Vibration is impetus flowing directly from the source to initiate impact with a perceiver. It produces sensations you identify as light, color, and heat. Any response in you is the result of experiencing sensation as a change in your condition.

"Alignment is obvious. Whatever you consider original light vibration, wave or particle, the components of that vibration arrange themselves into the wheel's directions and circular form. If at first glance, some element appears randomly unconnected, concentration reveals a larger field of vision in which that point is indeed aligned in some configuration with other points. Your universe is a web and nothing exists outside that web.

"Movement from one point to another within or around the wheel is the motion called wind. Vibration's first outward radiation initiates this movement. When it reaches the wheel's outer limits, wind extends its existence by circling. As the aligned elements tumble forward in wind's path, continual movement results. Momentum builds and wind endures as a self-sustaining force.

"That the combined elements of vibration, alignment, and

wind assemble a meaningful thought is evidence of order, the fourth element. Even if you cannot express that thought, you sense its presence. At some inner level, the mind recognizes order in whatever has been perceived. This inner level needs to awaken, or perhaps reawaken, before we can sustain the direct vision of the ancient heart. Otherwise, direct perception can drive you insane."

Wolf's words reminded me of a radio interview I'd heard with a Native American poet. She had been born on a reservation, educated in the dominant culture, and now was a university professor. She told of inner conflict resulting from her pursuit of success while whispered voices urged observance of wildly different values.

"I was successful," she said, "but I was terribly unhappy. I was getting two drastically different messages from my life and my heart. I had to go back home to the reservation to realize I was not insane, I was Indian. From my childhood I had learned to know by being in touch with my self, my source. As a young person in the world of formal education, I was taught the opposite. I learned to cut myself off from inner knowing and accept outside pursuits as the process of knowledge and reward. I did not remember my childhood feelings well enough to keep me grounded. I honestly thought I was going insane. When I returned to the reservation, I returned to the quiet ways. I talked to the old people and remembered the traditions I learned as a child. It healed me. I found out I was one with Mother Earth. I was not insane."

Whereas the poet recaptured her true identity to stave off insanity, I needed to release my sanity to envision identity. This process of release became my vision of the ancient heart, my destination in Twenty Count. An alienated society of the intellect harmed the child within the poet, made her feel insane. That society impacts many of us the same way.

"That happens," said Wolf, "because society structures its teachings on the fantasy road. This is the pathway of misperceptions. So often, adults fail to test out their own ideas about the world before passing them on to children. It is a sacred priority that nothing shall harm the children. If you are harmed as a child, then you must heal. This happens as you travel your Twenty Count path. The main thing you must bring to the path is patience. You have traveled fantasy road a long time. You must understand that time is a commodity to appreciate and befriend. You must use it well. Remember, it's a long count to Twenty."

I had to take Wolf's word for it. Twenty Count was not something I could read up on. Nor was it a matter of faith or any type of belief system. Just the opposite. It was a technique for understanding what's going on. True, I had to develop inner sight, but I also had to think clearly.

The Count's essence was still hidden in those days, perhaps in the same blind spot as my vision of the ancient heart. If so, it was secreted a great distance below my mind's surface noise. That can be a tough place to reach. To quiet and open the mind is, of course, a premier goal of any mystery quest. I preferred to pursue that quest through the Count rather than any alternative system.

Having observed friends pursuing various paths for years, I was–and still am–leery of many approaches to questing. One common way is to turn over self-responsibility to another human being, someone whose training has qualified him or her for the title "master." You surrender personal power, theoretically for a brief time only, in exchange for guidance past the mind noise, a sure-fire way to discover inner peace. What many a master fails to explain in advance is that this quest can be a gamble worthy of the lights of Las Vegas. You bet all your chips before identifying the wheel as roulette, not medicine.

So tempting is the promise of a quiet mind, however, that seekers line up to endure whatever is demanded. In essence, they say, "I can't do it myself. I've tried and failed. But I know this is worthwhile. So here, take me. Do with me whatever you wish. Just teach me, lead me. I'm yours." It's one thing to whisper this to your own inner guidance, quite another thing to say it to a master. Inner guidance, however unfathomable its teachings may be at first, will definitely have your best interest at heart.

I had no doubt there were many wonderful and honorable teachers somewhere out there. I had learned a lot from Red Thunder Cloud, among others. But at that point in my life, I felt substantially more grounded with Wolf and the slowly developing inner workings of Twenty Count than with any other possibility.

"Working with the questing mind, whether your own or someone else's, requires a particular perception of the tricky essence of mentality," Wolf warned. "This perception is not acquired automatically through the quest any more than through years of university study. Both quest and study may inspire you, but you cannot know for certain in advance about any master."

As all teachings paraphrase: Know thyself. When this eternally demanding command reappears, the seeker begins pulling out hair. A circle with no exit. If the mind is full of chatter, and you want to choose a way to eliminate the chatter, how can you quiet the mind sufficiently to choose that way? In other words, what the hell is this quest and what is this devil of a mind barrier? And what is this special perception required to work with a questing mind?

My quest for clarity narrowed to one question. Simply put, what did Wolf know that I didn't?

Wolf is often cryptic, always simple, and ruthlessly direct, warning me often not to ask questions unless I truly want and can handle the answers.

"I have an advantage over you," Wolf explained. "That is why I can teach you. As a human, your blessing and burden are one: an individuated mind. Think of the entire essence of human consciousness as a vast ocean. You are one drop of salty water, flipped up and out of that ocean for a moment. As you rise into the atmosphere, you realize that you possess awareness, your own individual system of perception. That's the blessing, a gift so amazing in its possibilities for feeling and thinking and acting that the desire to explore your entire range instantaneously inundates you as a burning obsession. The burden of individuated mind is that this obsession cannot be satisfied. It will drive and control you forever on your flight through the atmosphere. In time, you will evaporate into air molecules. There's no way out, no way for you to simply return to existing as part of the greater and much more incredible whole, your original ocean. No way, unless you passionately desire such a way and then defy your obsessions and create that way."

"OK, I get what you mean," I said. "But what is the advantage you have over me?"

"I am not separated from my whole. I simply am my whole. I am Wolf, one with everything known as Wolf, and your seeing me as a single being does not change that. So long as you limit your identity to the single human that you unquestionably are, you can never achieve the direct connection to the source that I just naturally enjoy. That is why you need me. I'm your role model for connection."

"And instead of having my own connection, I'm stuck with being obsessed with infinite possibilities?"

"Absolutely. You're dizzy with perceptions. All around the

central core of your individuated mind, which strangely enough is also the heart of your devil barrier, you've constructed massive walls of mythology, your story about your misery. And your misery has become the guardian of your mystery."

"But if I'm dizzy with possibilities, why is my story about misery?"

"Because you also believe that you have virtually no chance of achieving those possibilities. You're convinced of your alienation from your perceptions and your chances for happiness or achievement. You see both yourself and those concepts as solid entities with absolute borders and you think that they are far, far away from you. Beyond your grasp. This causes you mental agony, suffering, and pain, just like in a civil suit. You believe it best to face your limitations, not avoid them. It does not enter your mind that the limitations are actually fantasies. Avoidance and denial reign as the queen and king of the shadow world. But your avoidance and denial are not about your true limitations. They're about your fantasies. Your only chance is to cultivate rebellion and ride it to its ultimately liberating conclusion."

Wolf explained that humans get stuck in this condition and are virtually unable to rise above it. Humans share a universal inventory of frailties which serves as our catalogue of complaints, limitations, excuses, and alibis. We experience our frailties, live with them, rely on them, profit from them, and indulge in them. Thinking about ultimate, unlimited, and seemingly unattainable possibilities becomes an exercise in utter futility and frustration. Daydreaming forms as escapist fantasy, the product of the illusion of alienation.

"It's true you feel isolated," Wolf said. "Alone, your own presence is all you are sure of. To know any other presence, you must interpret your sensations. This means to see and then speculate on what you've seen. Nothing else offers you information. Unless you interpret your input, you don't know what you've seen. You become convinced you're all alone and also lonely. Every iota of lack of thinking on your part increases your impression of solitude."

"Explain to me about interpreting," I said. "It sounds like the root of all fantasy."

"It's merely your way of making sense out of vibration, alignment, wind, and order. If you hear a sound, you immediately remember that it had both a start and a stop. After the stop, you retain knowledge of that sound only if you recall its presence between start and stop. All perception works the same way. Whether you take in sensation through your eyes, ears, nose, tongue, or skin, you remember the experience by recalling its start and stop.

"All events of your life fall under this same umbrella. To speculate effectively on what's happening to you, you must isolate each event and its particular sensations. Unfortunately, the disciplined concentration needed for this is not the normal focus of most people. Instead, they prefer to think of life as a slightly hazy, fleeting thing they are powerless to comprehend. They welcome the crashing darkness of nightly sleep so their responsible time can end until the next light of day. An enormous portion of perceptive intake is dulled over by sleep and by fantasy, the dreamtime and daylight versions of toning down perception. But it's just as well. Direct perception can be disabling."

"But if I cannot endure direct perception, how can I reach clarity?"

"You will reach clarity within life's time. I know, because I am clear. And that's because, again, I'm complete, one with my all, and I don't need individuated mind to figure stuff out. As a result, I know something you don't. I know the way to beat your devil."

"What are you talking about?"

"You have a devil. Your mind. Remember, you live on fantasy road. Your only resolution opens through a simple realization: if mind is devil, then heart is friend. That which hides from the ploys of logic reveals its designs to the strings of feeling on the spiral path. When nothing else works, employ feeling, that deepest inner sensing, to confront the devil. That same sense recognizes order in the vision of the ancient heart. Not only know thyself, but trust thyself. Live from the heart. All you need now is to determine what heart is. And how to trust it."

"All right. What is heart and how do I trust it?"

"I thought you'd never ask. Heart and mind are one and the same, yet also very different. Let me shed some light into your mystery wheel. There's someone you need to meet."

MAP OF HIDDEN SECRETS

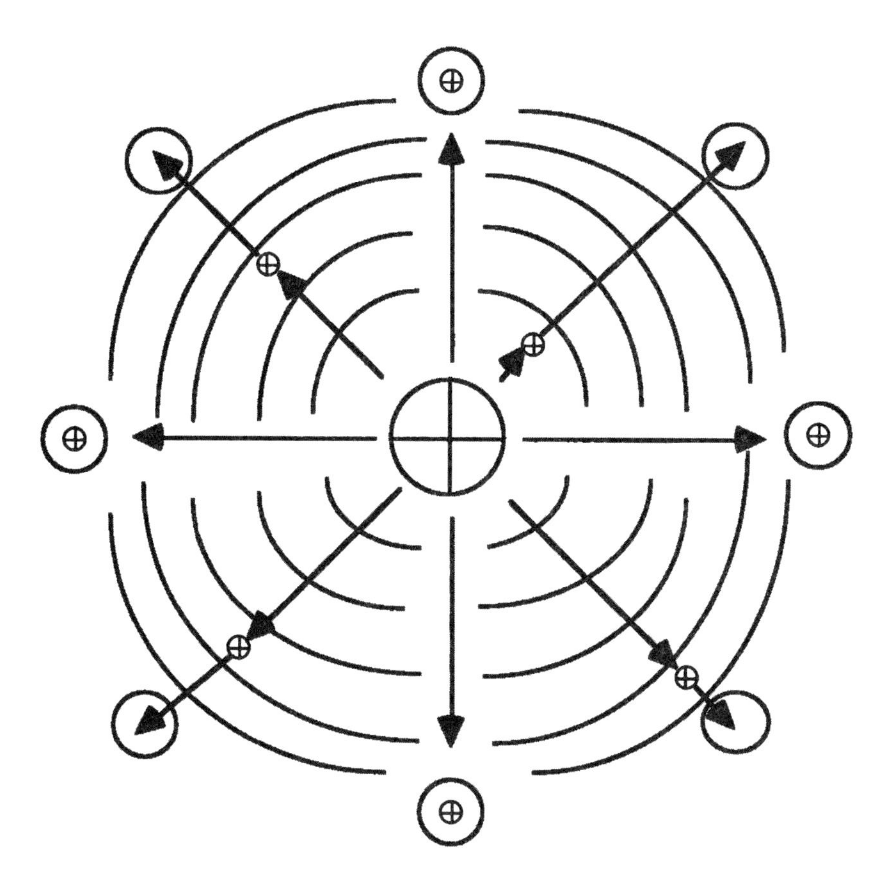

From source
in heart
to sensation
in mind.

Keetoowah's Gift

Keetoowah, a Cherokee shaman, struggled through the autumn of his seventy years in agony from smashed vertebrae sustained years earlier in an industrial accident. His little house in Santa Rosa, California, bore the untidy traits of a partially disabled man living alone, his wife already gone and no one to clean regularly. Young people visited and brought him marijuana to absorb the pain. He rasped complaints about his condition, spicing his monologues with humor and profanity. He told astounding tales of his life and travels, some in this world, others in far-distant realms. He spoke of energy and secrets and unicorns. He gifted visitors with advice and crystals. For his many years as a crystal master, he was known to students and other elders as the Crystal Godfather.

"Never surrender your thinking," the wiry, gnarled teacher commanded. "Change it, OK. But never just give it away. Unless you use your thinking, there's nothing anybody can teach you. To learn 'no-thing,' you have to use your brain power as well as you possibly can. Use your mind to understand you're not your mind, just like you're not your body. I'm an old man now, and it's hard to get around. If I really thought I was just this body, I'd be in trouble. This life has got damned hard for me. But I'm not this body. Do you know what I am? Do you know what we all are? Watch this. I'll show you exactly what we are."

Keetoowah aimed a flashlight at the wall, clicked on the beam, and declared, "See that? That little spot of light? That's all we are. That's me. That's you. That's exactly what we look like. We're the light shooting out from the heart of all things. If you'd spent a lifetime looking into crystals like I have, you'd know that too. You could damned well see it. I can see it because I live in alpha state all the time. What? What? You got a question?"

"I'm interested in a teaching tool called Twenty Count," I said. "I think it's based on old Maya and Aztec wisdom. Do you teach that?"

Wolf's guidance and my promise to deliver Red Thunder Cloud's sassafras gift had combined to lead me to Santa Rosa and Keetoowah. He fondled the bag of sassafras in one hand while holding a small crystal wand in the other. He did not speak for several minutes.

"The Maya knew a lot," he said. "They knew about the light. They knew about Zero, you know. That's the same thing as the light. It's nothing. Light is nothing. You can't touch it. You can't eat it. You can't rub it up against you. How do you think they ever got the idea for writing it down? It's nothing. No-thing. That was a good question you asked."

For the aging healer, the logic of his answer was impeccable. He spoke from the heart and his light was aimed at my mind.

Wolf prodded me into meditating on Keetoowah's flashlight beam image, and I exhausted many batteries during the following months as essential teachings shone on a wall in front of my eyes. The light beam's circular image of mind layered itself over my mental image of the mystery wheel. Wolf labeled this my memory map of hidden secrets, a picture of heart and mind as separate yet united.

In time, I managed to make out individual light vibrations radiating outward from the center of the beam into eight rays that merged into a circular vision. The rays of the four cardinal directions shone clearly and brightly; the four ordinal rays of southeast, southwest, northwest, and northeast, however, shone considerably less intensely. The cardinal rays divided my perception into quadrants of brightness interspersed with dullness.

Light was flowing from source to reality just as awareness travels from source to mind. The cardinal rays were predictable, always reaching the rim with full force, producing the combined sensations we interpret as ordinary reality. But a host of sensations from the ordinal rays landed seemingly at random in isolated nooks and crannies within the uncharted, whirling spiral of inner mind, not immediately available to surface perception. This was nonordinary reality.

In other words, my everyday awareness could take in only some of the elements of awareness flowing into my range of perception. The rest lodged as unconscious secrets deep within the many layers and folds of mind. My internal mental maelstrom further obscured these hidden entities like a sandstorm burying diamonds in a desert. These hidden mental entities are humanity's occult knowledge, the sought-after secrets of the ages.

My quest clarified. By unearthing these bits and pieces of

mental sensations from within my mind and interpreting them into a meaningful whole, I could recover my own inner knowing. I sensed a deeply obscured order in the light beam and in my perceptions and thoughts. Though I lacked power to see the big picture, the awareness flow from heart to mind projected its own unique logic. Heart was the start and mind was the end of sensation. Just as Wolf explained, I remembered sensation because I could perceive its beginning and end. It was all one thing, and so, strangely, was heart and mind. Just as the light beam's start and finish was one. A teaching was surfacing in me. I was getting it. Keetoowah really did know a deep secret about his flashlight beam. I said as much one day to a first-time visitor to the Grandfather.

"But if Keetoowah knows so much, if he is really a medicine man," asked the visitor, "then why does he suffer so? It doesn't make sense. He's in pain all the time, day and night. Why doesn't he heal himself?"

Between the innocence of that question and the complexities of any attempt to answer stands the barrier of human mystery–our lifelong enemy, nemesis, and challenger. To search the outer world for clues leads to the conclusion that resolution lies within. Delving within exposes layers and walls, questions and puzzles, fear and rage. Dead ends abound. We need clues, a map, but from where and to where? To comprehend our being, we must utilize reference points from the outside world, lest we end up communicating with only our own obsessed selves. Yet the easiest and most tempting trap is to accept someone else's packaged prescriptions and descriptions. Attacking barriers vicariously is easy, but ultimately pointless. In each of us, the mystery must grow into "my story."

Keetoowah, an elder of the Cherokee people, faced the mystery head-on, just as did Yogananda and Joan of Arc and Einstein and uncounted waking others. This Grandfather lived his truth and transcended his mystery. When he spoke of journeys to outer planets, of misplacing entire Earth years, of learning at the feet of unseen masters, he voiced his inner metaphor of direct truth. Clarity and trust in the invisible world provided the crux of his teachings, yet he constantly cautioned seekers against blind acceptance of inner voices.

"Being dead ain't the same as being smart," he lectured. "You can hear all kinds of voices in your head. Just because a spirit's not in a body doesn't mean it knows what's what. They'll tell you any damned thing. Just the same as you can see anything too. You have to learn the difference. You got to understand."

That was Keetoowah's credo: the old "know thyself" warning. Every teacher adopts it in some form as part of the mystery game. So simple a truth, so complex a command. Enormous power awaits in clear perception of simplicity. This is the teaching of the Zen story about the young disciple who, after hours of meditation, shouted to his master, "Roshi, Roshi, I've reached enlightenment! My eyes are closed and my breath is flowing and I've received the vision of the great, beautiful, golden Buddha covered with shining jewels and smiling in complete peace!" To which the master replied, "That's all right, my boy. Just keep breathing. It will go away."

By the time we're ready to investigate the simple truth of whether inner voices and shining visions may be less reliable than our auto brakes, we've usually long abandoned the majority position of knowledge by mass consensus. We appreciate blessings of the material world while acknowledging its spiritual lack.

We yearn for the essence of spirit but also realize that the most heartfelt wishing doesn't yield change in either inner or outer environment. We want to find something of substance and we're susceptible to being misled. Crusty old Keetoowah understood this.

As a tribal elder, Keetoowah was afforded privileges such as entering a sweat lodge in a counter-clockwise direction and wearing moccasins during traditionally barefoot ceremonies. While visiting Hawaii to study cancer cures in the Huna tradition, he was inducted by local kahunas into their number, not for what he learned from them but because of his life's accumulated wisdom, was the same as theirs. Knowledge does not exist in a vacuum. A focused, shared Earth-based wisdom flourishes, acknowledged and embraced by native peoples throughout the world. An ancient, clear, and infinite perception conditions this view of reality. Within this view resides guidance, obscure and obvious at once, to our questions about Keetoowah's pain and about our own inner pathways of seeking.

Searching for this worldview, we are moved to ask: What was the Crystal Godfather saying in his stories of outer orbits and inner teachings? What is the special wisdom that Native American spiritualists have handed down for generations and now are offering openly as our century ebbs away? What do Tibet's displaced lamas understand that eludes the world's teeming millions? What inspired the writings of Herman Hesse and Lao-tzu? What drives each of us to wonder about these questions and where do answers hide–and why? If Keetoowah knew so much, why couldn't he heal himself?

Depending on our natures and conditions and influences, we face or avoid confrontation with this mystery barrier every moment of Earth life. So long as we shy away—that is, deny responsibility for our own lives as well as all other life around us—we live in a condition described by various thinkers as the profane dream, sleep, denial, avoidance, the dark dance. But merely declaring that we want to confront the barrier is no answer either. Signing up with a teacher is no sure thing. As Keetoowah knew, false prophets thrive both within and without to lead us astray. Everybody has a mythology to sell.

"Mystery breeds mythology," Wolf confirmed. "When mind cannot sufficiently encompass the entire scope of a challenge, that which lands outside mental limits becomes the unknown. The quest to expand mind to include that unknown produces a technique of observing and explaining successes, gains, losses, and failures along the way. The unknown is the mystery, the technique is the mythology."

To enlighten mystery, the pathway of mind expansion must traverse and transcend mythology. The catch is that an effective pathway must be unique. Originality counts because mystery learns from everything that has gone before. Mystery has the most efficient computer in the cosmos and incessantly updates its program on the most recent successful effort to transcend humanity's mythology. Determined to prolong its own existence, mystery instantly devises a new defense against that recent transcendental success. To beat mystery, you've got to beat its system, computer and all. Creativity of the highest order is demanded.

Since mystery breeds mythology and, in turn, mythology resolves mystery, the key to creativity obviously dwells within the original mystery. Inside the inscrutable structure of Zen koans, those pithy but unanswerable questions asked by masters of their students. Inside Tibetan mandalas and Stone Age wheels and the yogic aphorisms of Patanjali. If we could get inside, we could understand, but understanding is the only way in. Since humans walk individual paths, each must devise an individual entry into mystery. That creative entryway is the goal of Twenty Count.

Through conversations with Wolf, I had come to understand that pursuit of the infinite ultimately involves projecting a system of circular reasoning and then accessing creativity to escape that circle by spiraling inward in search of a goal. However vague, however confusing, this was my starting point. To investigate the inward goal, I had to pursue a spiral. Fortunately, I knew firsthand where to find one.

A childhood dream had long imprinted my inner vision: I am a little boy holding a bow and arrow, sitting on a woman's knee. I shoot the arrow straight out at myself as dreamwatcher, and it spins into a black-and-white spiral, whirling and swirling as it approaches and engulfs dreamwatcher. The spiral, after apparently reaching some unknown realm beyond dreamwatcher, reverses its spin and fades back into little-boy-me again. Deliberately and with no emotion, I take up the bow and fire another arrow. The sequence repeats endlessly. The dream reappeared on a regular basis from early childhood through adolescence, and sporadically thereafter.

One night not long before Algonquin Peak, my friend James and I sat drinking plum wine in a Japanese restaurant in Providence. I related the dream to him. A Gestalt therapist from Scotland, James had a relaxed way with both wine and wisdom. He took a long sip of the former and said, "Sounds like the classic spiral of life. You know, the old idea that you are part of the spiral flowing round and round until you awaken and, instead of circling in repetition, you break free and create your own reality. Have you tried extending or finishing the dream?"

"I don't have that kind of control over my dreams," I said. "I can't just summon it up and then impose my will on it."

"That's not exactly what I mean," he said. "Since this dream came out of your own imagination, if there's more to it, then it will be right there waiting for you. Consult your imagination and see what's there. You meditate so it shouldn't be that difficult. Just picture the dream and when the spiral goes over your head, for instance, turn around and see where it goes. Maybe it's trying to show you something and your conscious mind is afraid to look behind you into the other side."

Whether it was James' advice or the effects of the plum wine, the dream returned that very night. When I woke and remembered the spiral, I followed his advice. Meditating as dreamwatcher, I activated my vision of the boy and the bow. I watched the arrow's flight as it flared into a spiraling ring aimed directly at me. As the spiral encompassed and virtually swallowed me, I pivoted my head to maintain vision of the spiral's speeding form. A flashing image lit up my inner eyelids. I maintained the view for only a moment. Then I followed the spiral image back to its boy and bow. I never again dreamed the spiral, but I meditated on it daily.

In time, I learned to hold the overall image of the new field of vision long enough to apprehend its construction and message. What I found was my personal entry point into the mystery: Twenty Count's cycle of teachings. Within the Count dwelt limitless possibilities of comprehension balanced on one thin line, a razor's edge of reality. This haziest of trails amid a world of wild imaginings offered teachings that the wise of all ages have translated into myth. To travel the trail was to slowly but surely unearth elements of the original light rays that did not make it to the mind's surface.

At that point, the mystery began to slowly unravel into a spinning wheel of wisdom. As new information surfaced from hidden mental recesses, Wolf delivered an introduction for lessons to come.

"There is a way to live that has been given to the people here. To fully understand that way is to enter the mystery barrier of mind. To just perceive that the mystery exists is the first tiny step toward that understanding. Be easy, go slow like a turtle, but get it. Only by proceeding through Twenty Count's intricate inner workings will your mystery turn into mastery. Then, in the tradition of the discipline, if you wish to keep learning, you must pass on those teachings."

As for Keetoowah healing his own injured spine, he himself provided the best answer, "You got to use your head to figure that out," he said more than once. "For some things, you just have to go to no-thing for your medicine. That's no-thing, you understand? Not nothing, but no-thing. It's where the light comes from, and where it goes back to. You got to use your damned head to understand that. Get it?"

Everyone who sought out the old healer was welcomed into his wisdom, some in puzzlement and some in understanding, but all in confrontation with the barrier of mind. His guidance surfaced in many forms.

"You say you want to know Twenty Count?" Keetoowah asked me one day, months after I'd first posed the question and he'd not answered. "I think that's a powerful tool. But I don't teach it. There's probably lots of secrets in it. There's old medicine wheels on this continent been here since the Stone Age, and I think that's what Twenty Count's about. Sure, that's probably your way. But I don't teach that, I never have. I don't think I can teach you no-thing either. I'm probably not your teacher."

And in years to come, as he welcomed more seekers and raged in pain and told tales of travel and spirits, he never again mentioned Twenty Count.

Thank you, Grandfather Keetoowah, for what you did not teach me. You merely showed me the flash of the crystalline light beam, the no-thing and its hidden secrets, and I appreciate it.

REALITY REALM

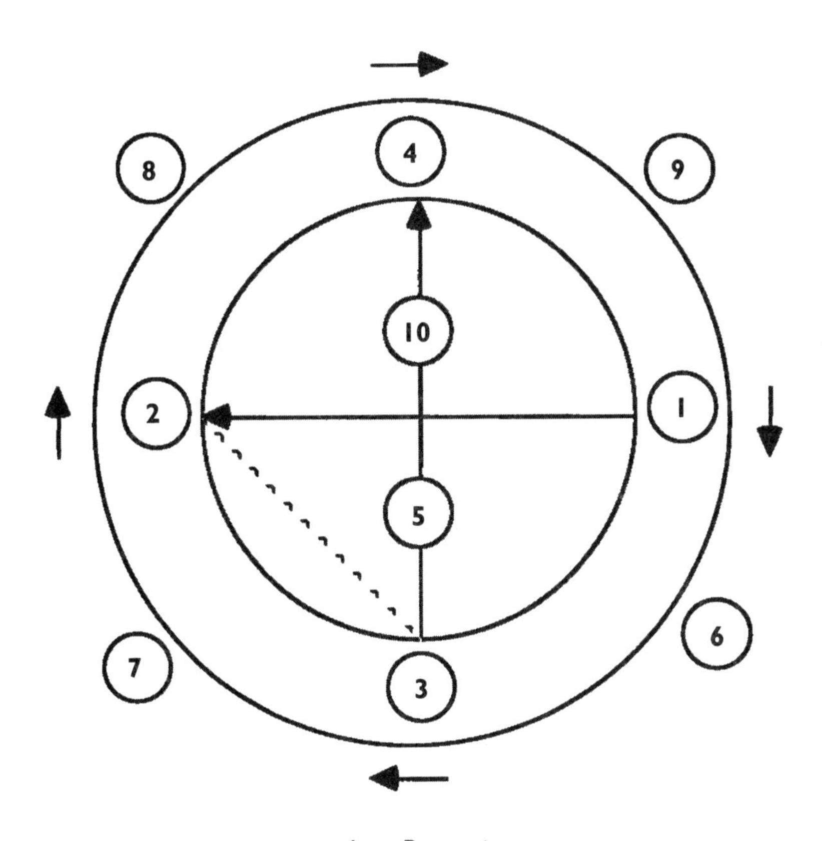

1 — Perception
2 — Reflection
3 — Projection
4 — Balance
5 — Separated Self
6 — History
7 — Dreams
8 — Relationships
9 — Choices
10 — Expanded Awareness

The Night with Comanche Jack

One autumn night in the California redwoods at Half Moon Bay, we gathered around a campfire as elders talked about medicine teachings. Red Hawk, Twylah Nitsch, and Sun Bear spoke. Wisdom flowed amongst the gentle tellers until nearly midnight.

When the elders retired, a few of us hunched close to the dwindling fire, speculating about the esoteric world of spirits, energy, and the like. People spoke like elders-in-training. But one by one, the seekers tired and went to their sleeping bags in the nearby woods. Finally, I sat alone with a skinny, tattooed Texan with long, blond braids and a handle-bar mustache. He called himself Comanche Jack. An aging hippie biker from the 1960s.

"You wanta see somethin'?" he drawled. "Somethin' you ain't never seen before?"

He picked up a scorched kindling stick and sharpened one end with his knife. Clearing some ground, he drew a large circle with a cross in the center and four large circles at the four directional points.

"Medicine wheel," I observed. Nothing special about that.

"Keep watchin'."

He continued drawing small circles, both inside and outside the original wheel. At the heart of the cross, he drew a sideways oval. Chills crept up my legs and body. He sketched a total of twenty small circles.

"That's Twenty Count," Comanche Jack said. "It's Cherokee, I think. Supposed to come from the Maya. An old woman taught me. Don't know where she got it, but ain't nothin' like it. Guaranteed."

"There's something like it," I said. "Not exactly, maybe not as sophisticated, but something like it." I pulled out my bag of arrowheads and showed him the painted wheels.

"Interestin'. You got teachin' for this?"

I told him about Keetoowah's no-thing, my map of hidden secrets, and the four aspects of Zero. He listened closely, smiling and nodding.

"That's good. I like it. Well, maybe I can give you some more. It'll fit in real smooth with what you already got."

We tossed wood on the fire and talked for hours. Comanche Jack related a teaching about each of the twenty circles, pointing to the circles one by one with his sharpened stick. As he spoke through the night, Wolf's voice chimed in simultaneously in my inner ear. I listened to both sets of comments on all twenty numbers.

Nothing since has reproduced the singular atmosphere of that night as the form and substance of Twenty Count came alive for me. I finally slept shortly before daybreak. Comanche Jack sped off hours later on a high-powered Harley, and I never saw him again. But he left behind an immense gift.

Here are the essential teachings of the numbers of Twenty Count as narrated by Comanche Jack, supplemented by a running commentary from Wolf. The first ten numbers deal with defining our primary reality, telling us who and what we are, and creating a vision of the mental nature of our universe. They offer guidance into deep memory on both an individual and collective basis. Each number builds on all that has gone before and also sets the stage for the numbers that follow. This is the reality realm.

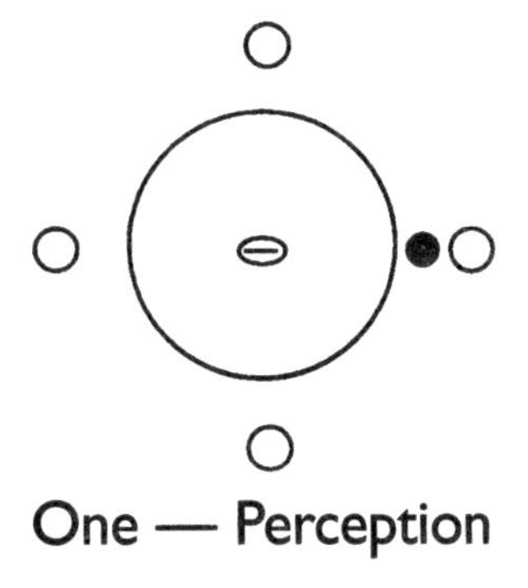

One — Perception

Jack: "This here's the startin' point for you and the universe too. Sun comes up, right? That's One. And you start out by gettin' zapped with light. Otherwise, you're just blank. Takin' in information is where your existence starts. Could be a smell, a taste, somethin' you hear or see. But it's input from somewhere out there. Or maybe in there."

Wolf: "Great. You finally found your master, your guru. Funny, he doesn't talk like a guru. But he's giving you the starting point and the landmarks along the road of the Twenty Count trek. So long as you walk a straight path, you can end up as Master of Twenty Count as well. Don't worry. You'll know when you arrive."

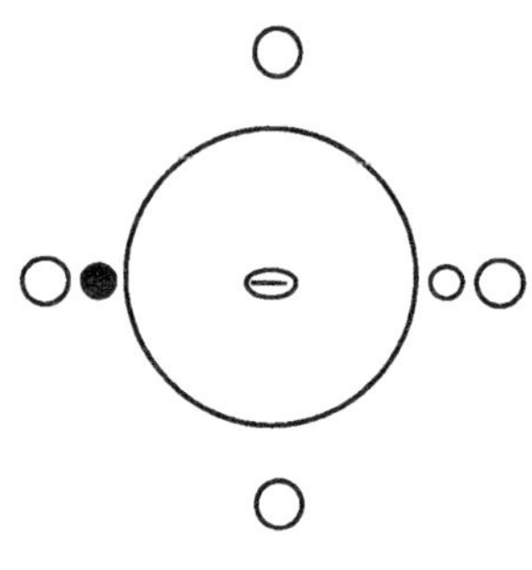

Two — Reflection

JACK: "Second step, you get a base, like good old Mother Earth, a place where stuff can grow. This means sit down and think about whatever input you got. Run it around inside your skull and see if you can figure out what it means. You know, does it taste good or maybe you should spit it out?"

WOLF: "Good point. This is the process of examining the nature of our world. It's good to be aware that lots of fantasies start right here, such as the fantasy that the material world is made up of solid matter. To see through this fantasy, simply put a slice of anything on earth under a microscope and turn up the power. Ultimately, you'll see nothing more than patterns of relationships. The light and shadow show."

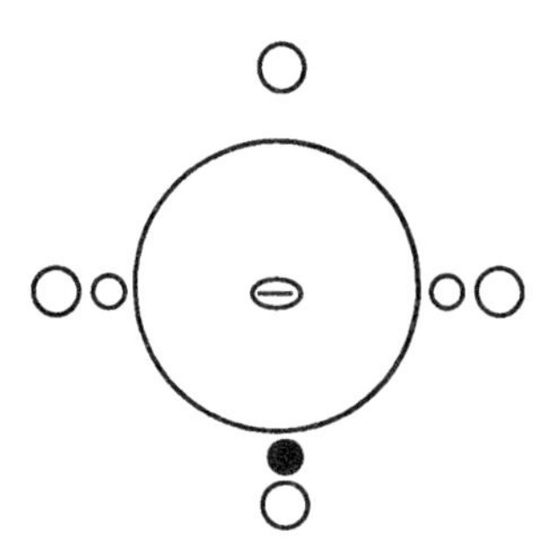

Three — Projection

JACK: "This here's where the green stuff grows. The plant people, all the good salad stuff and smokin' stuff. For us people, it means puttin' out, doin' somethin' with whatever we came up with while we were reflectin'."

WOLF: "Three's a major number because it initiates geometry. For instance, with three sides and three angles, you have a triangle. With a triangle, you have a basis for building structures on your planet."

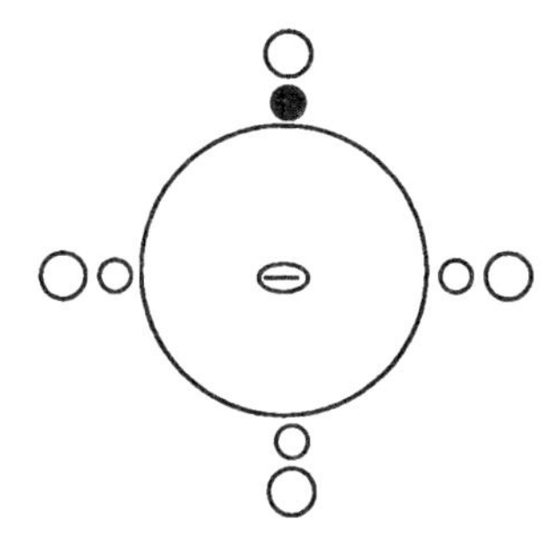

Four — Balance

JACK: "This one's the animals, all the fliers and buzzers and four-leggeds and whatever. See, the idea's that the animals get nourished by the plants that grow out of the ground that got formed by some big bang that made the sun way back when. There's an order to all this. Four means you got to stand upright in your human skin and bones and walk animal-like."

WOLF: "With four sides and four angles, you can make a square. Mathematics is growing, and so is your mental capacity. Remember from Comanche Jack's drawing that One was east, Two was west, Three was south, and Four is north. The significance is that if you connect these numbers in order, you've created a cross. Remember this until we get to Nine. It'll be significant then."

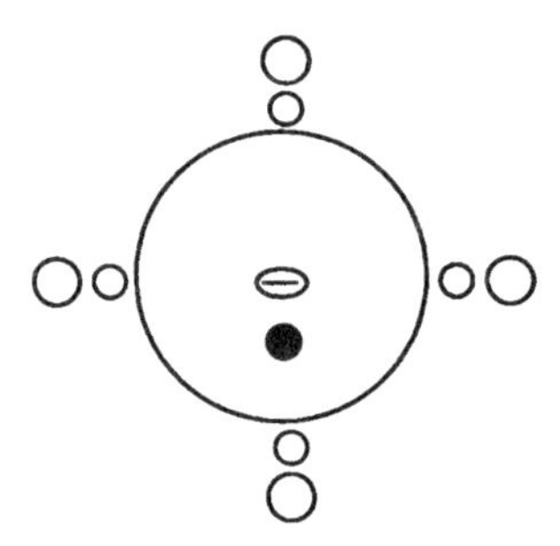

Five — Separated Self

JACK: "Now this's where humans pop up in the whole scheme of things. Just like the animals and the plants, they're children of Mother Earth and Father Sun. We're all here together, one family. This one shows up inside your head as bein' able to see the whole picture and thinkin' that you're the only one here who can think. Animals and plants and stones, they can't add and subtract and talk about it. You get to feelin' you're pretty damn special."

WOLF: "Talk about a place that generates fantasies, this is it. When you stand as a human on Earth and look around, a feeling of superiority hits you like a ton of ego. You get the feeling you're lord of the Earth. That you can outthink all other beings and therefore they are here to serve you. And that same principle can apply to other people as well. Be aware. Be aware."

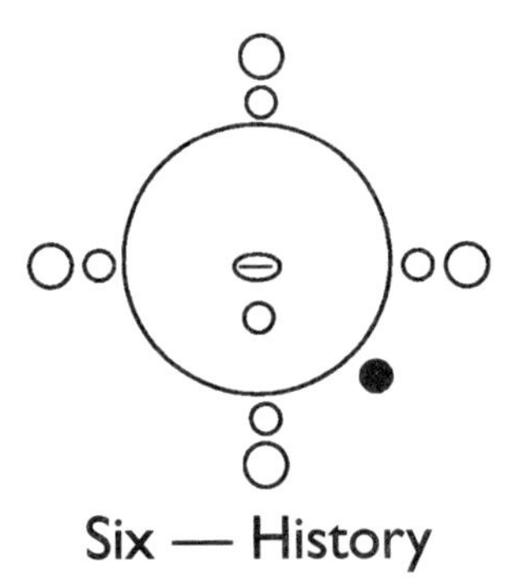

Six — History

JACK: "Now we're detourin' into the world inside your own little mind. Stuff we can't see. History's about all your past influences. Your genes and teachers and what not. Whatever's made you what you are. Thing is, you probably ain't never thought about half the things that've influenced how you think and act. Those influences are just layin' around in there, livin' off your energy, and blockin' you from freedom in your thoughts and action. This stuff gives you lots of bad feelin', obsessions and such. You got to get yourself clear."

WOLF: "Your environment, your education, your early life experiences, your physical and mental characteristics. All these contribute to your basic view of who you are. If you accept all these determining factors without questioning them, your life will pass in convincing yourself that your basic view is so absolutely correct that there is no point in trying to grow and change."

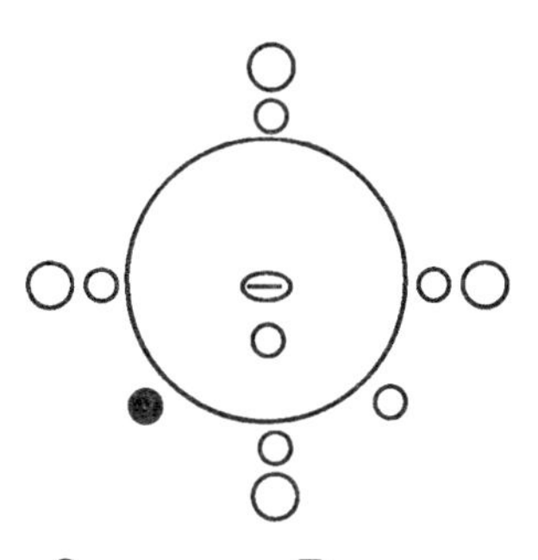

Seven — Dreams

JACK: "Yeah, the dream world. Lots of ways to get in here, not just sleepin', you know. Meditatin', good smoke, even booze. Thing is, there's as many nightmares as good dreams, and whatever we dream, that's what comes up in our real lives. Scary, ain't it. See, we got to get some control over our dreamin' or we're never gonna get anywhere in our lives."

WOLF: "This is the place of all phenomena. It's easy to get lost in here and to get caught up in the fantasy of having no control over your life. That, in turn, turns you into a person obsessed with trying to control everything."

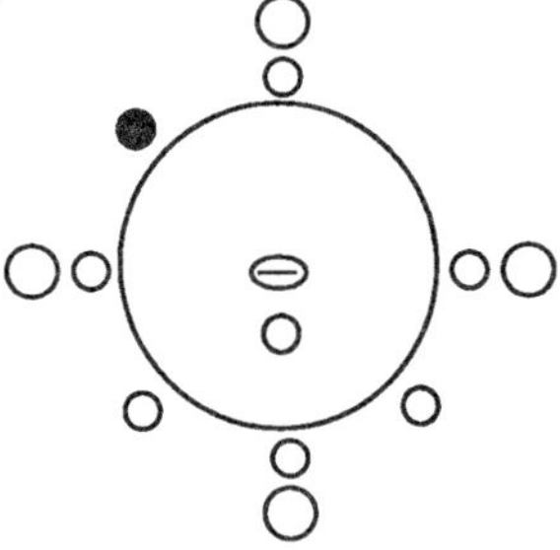

Eight — Relationships

JACK: "Karma, man, karma. What goes around comes around. How you treat everybody else, that's what you get back. If your attitudes toward other people are screwed up by your history and your dreams, then your actions ain't gonna be any better. Relation-ships are all about karma, you know, the ties that blind. You gotta treat people right. And plants and animals and Earth too."

WOLF: "If you want to control everything, you'll end up as the victim of the fantasy of holding yourself responsible. Anything that goes wrong on Earth can become your fault. You'll accumulate enough guilt and blame to offset all the praise and pride that you treasure so much."

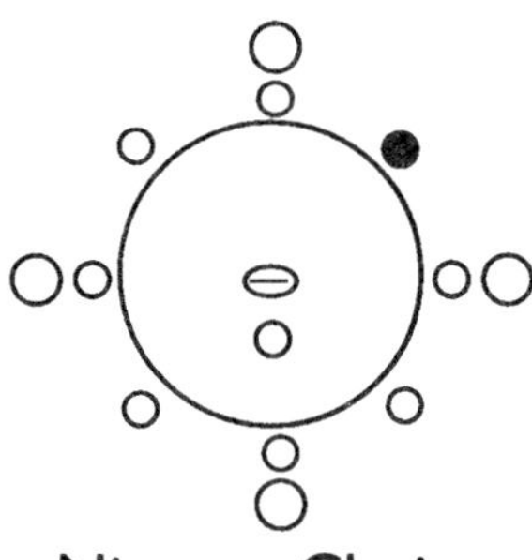

Nine — Choices

JACK: "Look at it this way. If you got a rotten history, nightmares, and lousy relationships, what chance you got to make good choices in your life? Most likely you'll choose somethin' that, time after time, just keeps you spinnin' around in rotten history, nightmares, lousy relationships, and more bad choices. Sooner or later, though, maybe in this life or the next or the next, you'll choose somethin' real good that'll open up your heart and head. Like findin' just the right lover at the right place at the right time."

WOLF: "Remember the cross we created with numbers One through Four. Now trace a circle from Six in southeast to Seven in southwest to Eight in northwest to Nine in northeast and then back to Six. Cross in a circle. It's your basic medicine wheel paradigm."

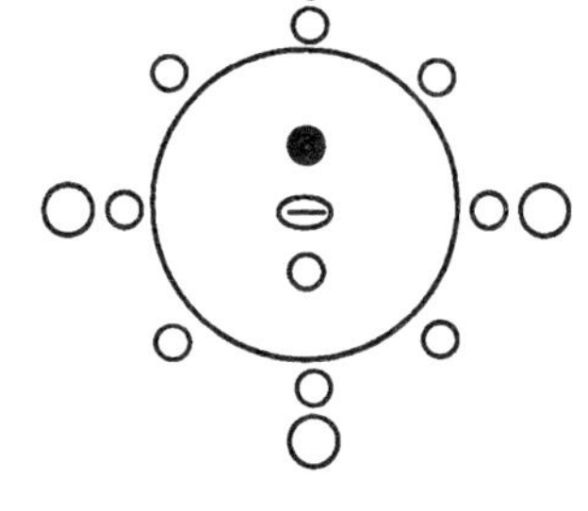

Ten — Expanded Awareness

JACK: "Whoa, now. This is a good one. This means wakin' up and seein' a bigger picture. You get religion at Ten. You realize that you're more than your upbringing and your scary dreams and failed relationships and bad choices. You realize there's a state of grace waitin' for you. This is a cool place. I always hear that old hymn here. You know, 'Amaaazin' grace, how sweeet thou art, that saaaved a wretch liiike meee . . .'"

WOLF: "Great. The music man. Well, he means well, but like a lot of folks, he loves to trip out on religion. If we see separated self at Five as the embodiment of humanity's tendencies toward superstition–the primitive individual trying to find the right tools to deal with worldly realities–then Ten is definitely the religious state. Glory to whatever god and whatever goddess, and sing their praises.

"Before getting too schmoozy with the angels, let's move on. We're ready to get into another level of numbers called the smoky unknown, and you're going to need all your grounding to find your pathway through."

ABSTRACT REALM

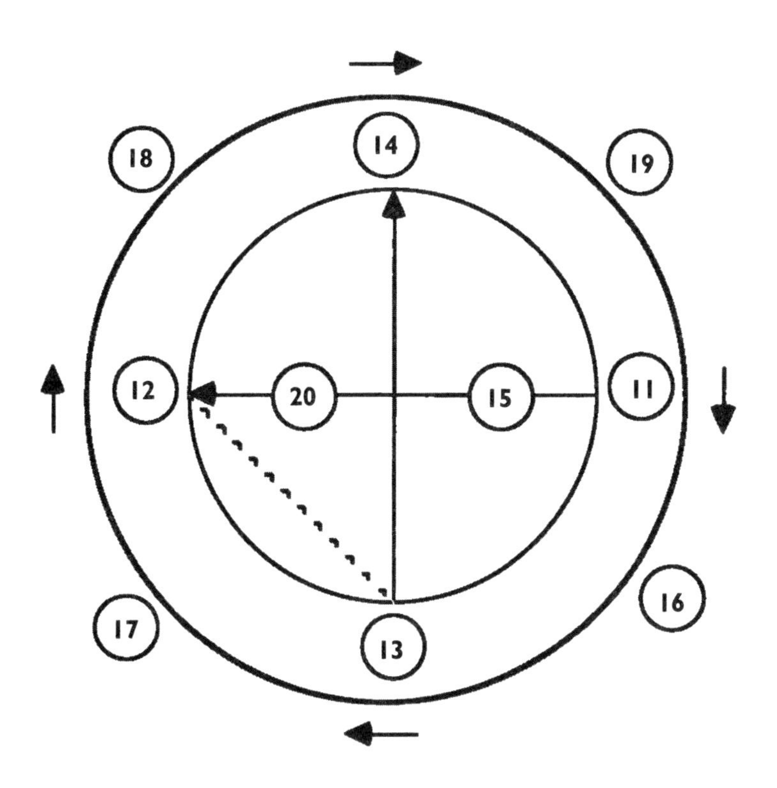

11 — Inspiration
12 — Organization
13 — Transformation
14 — Instinctual Intellect
15 — Kaleidoscopia
16 — Synchronicity
17 — Imagination
18 — Empowerment
19 — Freedom
20 — Return to Ancient Heart

Smoke Knowledge

The realms of Twenty Count are as different as night and day. They are outer and inner being, the known and unknown. One is world, one is smoke. The teachings of the smoky unknown are fleeting, like flowers you do not remember loving for their fragile beauty.

The only way to penetrate the elusive unknown is to see. In Native American symbology, smoke represents spirit made visible. When you see smoke rise into the sky, just keep watching. Do not turn away. In time, you will see the vastness into which the smoke extends.

Red Thunder Cloud crafted beautiful pipes for other people, but he did not smoke them himself. He simply could not stand the inhaling. He told a story about attending a televised gathering when a ceremonial pipe was passed around for everyone in a circle to smoke as required by tradition. When the pipe and camera reached him, Red Thunder Cloud put up a brave front, closed his eyes, inhaled, and doubled over in a coughing, choking fit while everyone else broke into laughter.

"I'm a medicine man who can't smoke a pipe," he moralized. "I just can't handle it. I have to take smoke on faith and knowing."

That's what numbers Eleven through Twenty are like. They have the shape and feel of something familiar, like a beautiful pipe we have created for smoking. But to prepare the way for this level of teachings, a powerful inner process must unite our waking and dreaming states and open our minds and hearts simultaneously. This process is both the method of access and the underlying message of the smoke knowledge.

Let us resume our study with Comanche Jack and Wolf. Jack continued his recital as the fire burned low, and Wolf's voice went on simultaneously in my inner ear.

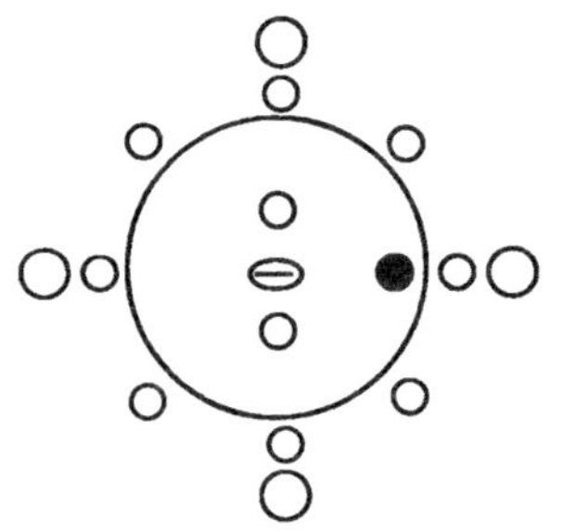

Eleven — Inspiration

Jack: "Now that you got religion at Ten, you start lookin' outward instead of inward. You see the stars shinin' and twinklin' up there and you think, 'Man, that's nice. That's inspirin'. You feel like you're one with God and you just love everybody and everything. Love, man, I say love."

Wolf: "He's got that loving feeling because he's added expanded awareness, Ten, to perception, One, to create Eleven. That's how this entire realm of numbers comes into being. It's all the result of expanded awareness. So beware, if you lose that expansion, your perception slips back into the reality realm."

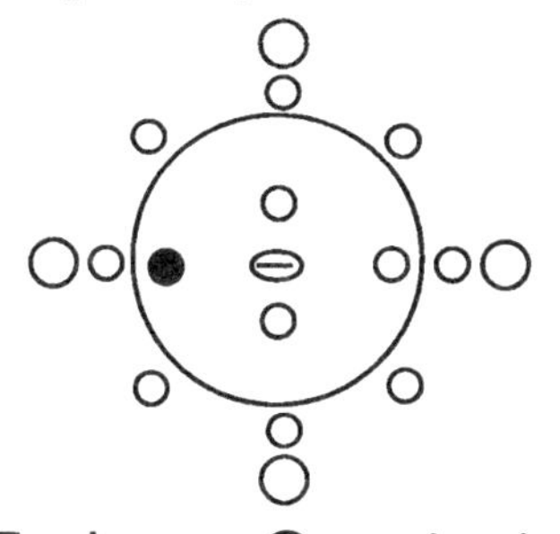

Twelve — Organization

JACK: "While you're gazin' up at the stars, you strain your eyeballs a little more and maybe even grab a telescope, and hey, there's the moon and Venus and all the other planets. They're all lined up in this system of orbits and rotations, and if you study them long enough, they make sense. There's this organized feel to the planets, and you get the feelin' that just maybe your own life is organized too. You just gotta figure out how to look at it and study it."

WOLF: "You see, Jack's saying that there's a big picture available here. When you start talking about cosmic organization, you're talking about self-based order. This organizational principle comes into being as you begin to see the material world as a fluid but unified field of aligned, orderly vibration."

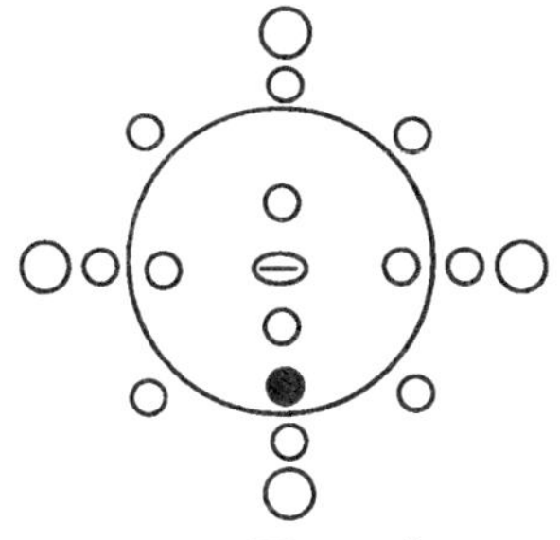

Thirteen — Transformation

JACK: "Thing is, now that you're all inspired and want to see how things really work, there's no way you can figure it out if you keep walkin' and talkin' with all your old reference points. Or maybe even with any of them. So, if you want to play, you gotta pay. This means dumpin' it all. Not on anybody else, you understand. Just dumpin' it. All your old history and nightmares and rotten relationships and lousy choices. And all the fantasies that came out of your obsessin' on those things. You gotta let go. Really let go. I call this the baby death. The needy little crybaby inside your mind must die."

WOLF: "He's right, and baby death is a good label for Thirteen. This is the place and time for transformation. Think about it. If you want to change and grow, you must do it with a different approach than the one that got you here. Here's where you make the change."

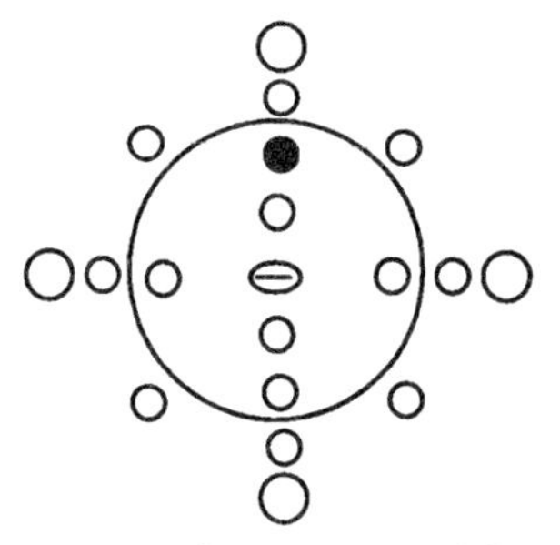

Fourteen — Instinctual Intellect

JACK: "When baby dies, somethin' else takes over. It just comes up and takes you over. One day you're one way and the next day you're another way. This thing is instinct, pure animal power. Not intuition, you know, that's just a weak little bleed-through of instinct into our everyday lives. But I'm talkin' here about knowin' naturally through the body how to act and how to talk the very best way you can in every situation. Just like animals know how to survive. When this instinct comes up in you, it kind of blends into your brain cells or somethin'. Then you got both intellect and instinct actin' together, and, hey man, it don't get much better."

Wolf: "The cross of the smoke teachings is now complete, same as before, east to west to south to north. Four directions in the higher realm. This means that your base for pursuit of wisdom is inspiration, organization, transformation, and instinctual wisdom. How can you go wrong?"

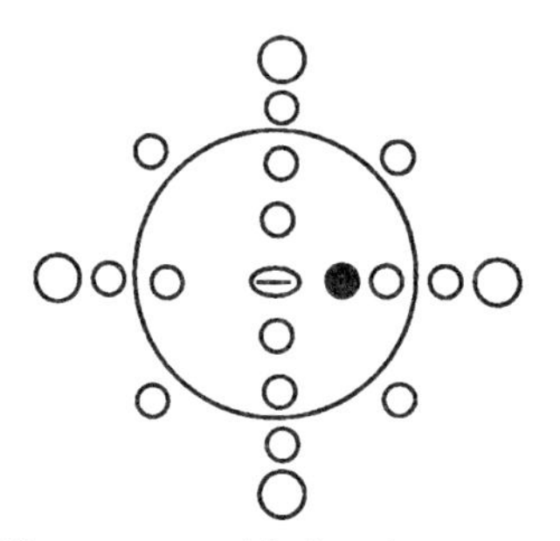

Fifteen — Kaleidoscopia

Jack: "I took some pills a bunch of years ago and went travelin' all over the universe one night. Then I felt lousy for months. My body couldn't accommodate the trip. I felt like I was in first class but it turned out I was down in the baggage compartment. Then I learned to do the same thing by meditatin'. No after-effects. Great. Now I know that when you get the old instinctual intellect fired up, this wild and wacky place I call 'kaleidoscopia' is right here on the inner horizon. This is all time, all space, all matter, all energy. All right here."

Wolf: "Holographic awareness. The commitment to religion at Ten jumps to spirituality at Fifteen as you realize that the cosmos can't be captured in a set of rules about what to do and what not to do. This is the realm of everything that is, was, and will be. It's here. Again, you just have to know the tricks of getting it and getting it right."

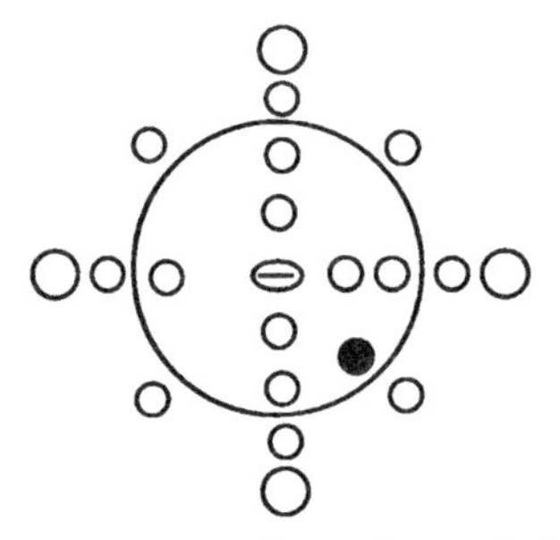

Sixteen — Synchronicity

JACK: "This here's a biggie. Once you dump the effects of your old history, you get to feelin' kind of strange. Like there's nothin' you can count on. You've given up your attachment to the straight-line, cause-and-effect chronology of life. You need a replacement base that you can think from. Well, that's synchronicity. You know, Jung's word for all the coincidences in our lives that aren't really coincidences. They're all really part of a flow that includes its own versions of inspiration, organization, transformation, and instinctual intellect."

WOLF: "This is the process of remembering that you are constantly forgetting. You are forgetting the deep inner messages of consciousness that are, in reality, as much a part of you as the skin on your elbows and the hairs in your nose. Once you learn to remember, you enter into a process of ease of movement in the outer world. Opportunities present themselves one after the other, all aiming you toward your own success. Just one thing—unless you keep on remembering, you won't know what that success will be until after it actually happens. In order to be fully informed, you can't let the remembering stop."

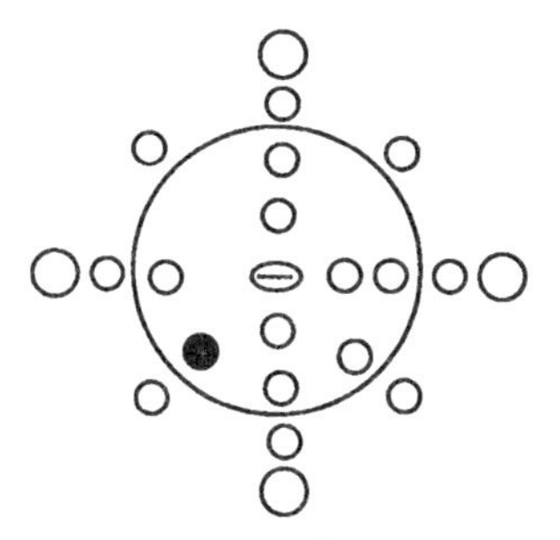

Seventeen — Imagination

JACK: "O.K. Now you got synchronicity comin' up in your life. More and more all the time now that you see the big picture of kaleidoscopia. But guess what. You better bring back a little of your old sense of chronology or you'll be ridin' your bike right off the Golden Gate first chance you get. There ain't nothin' wrong with chronology, you see. It's just that it's not the whole story. And neither is synchronicity. Put the two together and know what you got? Imagination, that's what. You see alternatives everywhere. No way you can ever be limited by your old history, dreams, relationships, and choices. And no way you can get lost in your new world of kaleidoscopia and synchronicity. You'll just get more and more creative. Cool, huh?"

WOLF: "Tuning into your imagination involves isolating events and perceptions in space and time. This is another part of remembering. It's also what any visual artist accomplishes through sculpting or painting. And it's precisely what the musician does all the time. The purer the note, the wilder the riff, the cleaner the isolation in space and time. You can do this with anything you do, anything at all."

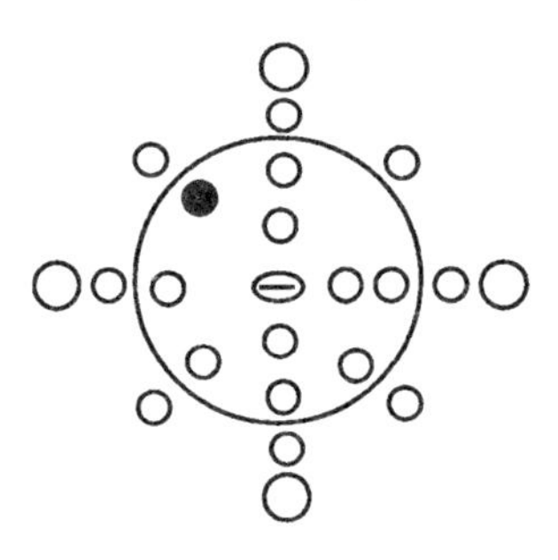

Eighteen — Empowerment

JACK: "Here's where it really gets good. Now you got a big picture view and imagination too. And here comes the full power of free will. You can do whatever you want. The old karma problem's been solved. No more ties that blind. You just focus your intent and start producin' anything you want. It's better to start out by workin' on nonmaterial things like your attitude and inner feelin', but stay with it and all kinds of stuff can happen. As long as you go straight ahead, no need to worry. You won't make any mistakes from this place and this system ain't designed to work if you try to create negative stuff for anybody, yourself included."

WOLF: "Want to be a success in business? Get a thought atom. Want a new lover? Thought atom. Want to write a book? Thought atom. Want to quit your job and strike out on your own? Thought atom. Want to understand what I'm talking about? Thought atom. Thought atoms are at the core of everything in existence. Everything you do, everything you say, everything you think–it all results from thought atoms and it all creates thought atoms. Want to know what a thought atom is? Twenty Count is a thought atom. A thought atom is an idea that you embody to such a degree that it manifests as your own intended purpose and action."

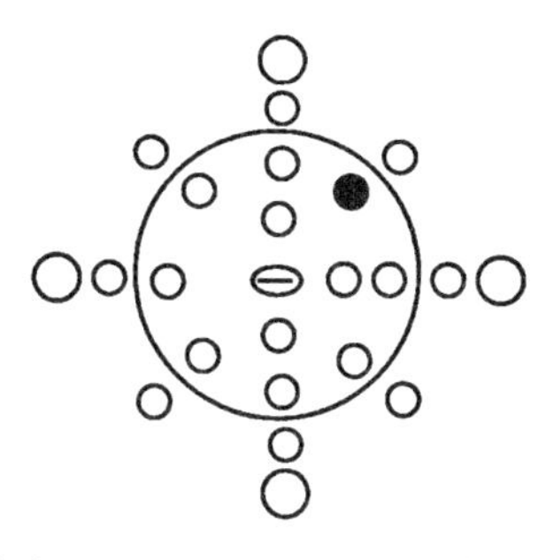

Nineteen — Freedom

JACK: "So here you be. You've done your work on yourself, raised your awareness, and you're livin' in the flow, bein' creative. What will you do with all this? Well, you don't even have to think about it. The answer's already there, and before you know it, somebody's gonna be whisperin' it to you and showin' you little visions of it and droppin' hints all over your life until you get it. And you know what? In some way, maybe obvious or maybe not, you're gonna be doin' somethin' good for other people and this planet and for yourself too. That's just the way it works."

WOLF: "Spirituality, the descendant of superstition and religion, transforms into freedom here. The wheel completes its ring of synchronicity, imagination, empowerment, and freedom."

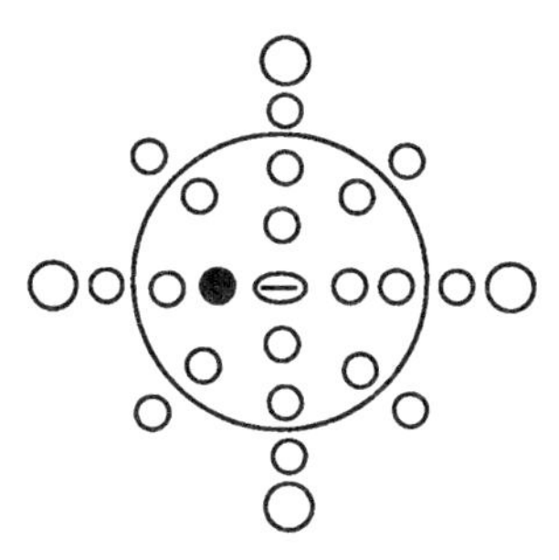

Twenty — Return to Ancient Heart

JACK: "Now this here is your final circle. End of the spiral, but not really. See, you've made it back to the creative center. Back where all these ideas and this whole process came from in the first place. And you think this little Twenty Count is all there is? No way, brother. See, what we want to do is get to the point of Twenty all the time. Then creation just acts through us continuously. Totally cool, man, totally cool."

WOLF: "Arrival at Twenty completes your journey, and the value of the experience gained goes into establishing the value of Zero as the source where you will renew yourself and then continue your travels. As I explained long ago, the vision of the ancient heart is both start and finish. This means that when you're ready, you must pass these teachings on, in your own way and your own words. That's how your journey continues.

"Now you know the turf the twenty spirits call home. But the bottom line still lies far ahead. So far, these 'spirits' are just concepts to you. For them to guide you into true clarity, you must allow them to actually come alive within you. You'll know when they do. You'll feel them like a shiver up your spine."

"Up my spine? Fine. But can you tell me one thing?" I asked. "There is a feel and a logic to everything I see here, but I did not either dream up or think up this entire, mind-boggling process. I couldn't have. So what is the origin of this teaching? Where did it come from? Why am I sitting here right now in the land of *déjà vu* talking to you about Twenty Count and twenty spirits? I thought Twenty would be more of an answer."

For a long moment, Wolf's violet eyes burned into mine. My mind blanked, and then I saw. Reaching Twenty was a means, not an end. I had to go through the passageway into the midst of the ancient heart itself.

My passageway came in a flash.

THE GOLDEN LIGHT OF COYOTE'S EYE

Looking at the other side of truth.

Primordial Flash

Dawn broke wet and chilled in mist. I pulled the sleeping bag up over my head and crept down into the warmth. Comanche Jack and Wolf and the campfire soared through my vision. A deep relaxation settled in as I recalled Jack speaking into one side of me and Wolf into the other. Sleep returned.

"You are spinning," Wolf said. "Spinning deep. Deeper than ever before. You must follow what I say or you will be lost. Spinning is all you feel, but the light keeps you from being dizzy. Look into the light. It is a friend. It is the singular light of soul. The whirling form of force. Music fills your ears. An orchestra of tones. Everything you've ever seen, ever heard, comes and goes. Appearance, disappearance. Light glows just inside your eyelids and you cannot forget the memory imprinted there. It is just before daybreak, you are in the depths of your dreams, Comanche Jack has gone, and this is the primordial flash. This is the moment."

The white light hovered omnipresent. At the center of the flash, the people sat cross-legged in rows of circles, some sleeping, some stirring, a few wide awake. They formed a primordial medicine wheel. Solar brothers and lunar sisters together. No one existed. No one worried. No Type-A personalities. Each was part of the whole, yet each unique. The flash endured only an instant, and each responded as best as memory and response allowed.

Suddenly the flash erupted into a blinding lightning bolt. The blaze struck brilliantly, roaring through on a wind so immediate that it vanished the same instant it appeared. Each in the circle gulped in a breath of wind, exhaled violently, and scrambled alive to pursue an image of lightning's finite essence. Into avenues of physical existence, they all lunged in pursuit of the most immediate creature of flight.

Fox, clever and swift, was the object of their chase. The pursuers were stumblebums. Instant by instant, Fox receded into its den of holographic memory until lingering only as a ghostly dot on mind's horizon. Undisciplined eyes lost the vision. Convinced the aspect had disappeared, they rejoined the circle as it assembled anew to wait for the next flash. Next time, each resolved, I'll get it. So difficult to remember the revelation, yet so impossible to forget the flash. Existence wiped out the tracks of light, but the path remained. Obviously, some traveler passed this way. All in the circle wondered if they were that traveler. Was I, each asked, once part of existence? Will I be again?

A lone being, however, held the vision of the ghostly dot, clung to the image, and doggedly followed Fox through the mist. The being ran, forgetting the tendency to stumble, and gained ground. The being leaped and grasped Fox's golden-red tail, unafraid of the clever one's teeth and craftiness. Fox did not slow down, just turned to nip the outmatched seeker. In time, the enduring being secured a hold on flying Fox and relaxed sufficiently to identify a bit of the kaleidoscope of life passing by. Reflections from the fleeting scene revealed the being's identity. I was the seeker, a rider on Fox in the wind.

My first impression of passing phenomena included whirling vibrations all around, each complete within itself, yet also touching and intermingling with all others. Gradually, my perception slowed the vibing. Colors and sounds began to align. The world was forming.

"They are liars, you know."

"Huh? What?"

"Liars."

"Who's there? Who's talking? Wolf?"

"Do I sound like Wolf?"

"Yes. No. Not really."

"Can you see me?"

"A little. You look like Wolf too. But no, not really."

It was not Wolf. The eyes reflected golden light, not violet. It was not Fox. The fur was the tawny brown of earth. And there was a grin. A strange grin. Then I knew.

"You are Coyote."

"Bingo."

Coyote, fabled teacher of American Indian cultures. Trickster, shapeshifter, sacred fool. And apparently my new dream guide.

"I said they're liars."

"Liars? Who?"

"Comanche Jack. And Wolf. Do you really believe all that garbage about something called Twenty Count? Do you really think there's an answer to this whole existence quiz? Just how gullible are you?"

In the instant of the asking, all satisfaction at finally finding the inner teachings of the Count vanished.

"I don't know," I stammered. "It felt so right. They were so together. It's what I've been preparing for. And that's why I'm here right now. I asked Wolf about the source of Twenty Count. And then this flash, and now you . . ."

"You get excited real easily, don't you?"Coyote grinned. "That flash, you know, was the singular light of soul. Don't worry, in a little while it'll vanish deep into the recesses of your memory. Tell me, what do you think is the value of learning to count from one to twenty and then attaching abstract meanings to each number? You could have done that on your fingers and toes when you were five years old if you'd had the smarts for it. Then what? Would you have applied for early admission to the eldership program?"

My deep consciousness laughed in spite of its apprehension. I reminded myself to remember this was Coyote speaking. No statement was necessarily what it seemed.

"You see," Coyote said, "none of their teachings is true. Do you know where you are at this moment? Do you recall how you got here? This is the primordial flash. This is Twenty. This is where you came from and where you've been trying to return all your life. This is the deepest level of awareness that you can possibly reach. Here, at this pinnacle of depth, I appear and tell you the truth. And the truth is you can't rely on a thing you've been taught about Twenty Count."

"All right," I conceded. "What can I rely on?"

"Me, of course."

"You?"

"I am the only creature, the only guide in existence, who knows the secret of living on your planet. I know how to handle your history and dreams and relationships and choices so that you will live happily ever after."

"That's quite a claim."

"It's true," Coyote stressed. "The secret is to let go of your pathway. Forget the teachings. You are taking it all too seriously. You see, this existence is a game. The mystery game. You're treating it like a Shakespearean tragedy, like a torture session with the Inquisition. As you travel through this life, most people you meet are obsessed with their own personal activities. Their jobs, their mates, their work, their fates. You left the workday world of career and respectability years ago, true enough. But you've brought the same serious approach to your seeking that you brought to your reporting work. You need to find a space in your existence where your mind can relax, free of dependency on the need to find answers and attachment to your vision of yourself as a free and clear human being. Only when your dependencies and attachments vanish will you be a free and clear answer man."

"All right," I said. "You've got my attention. Go on."

"It's all a matter of perspective," Coyote said. "Just as you were listening to Comanche Jack through one ear and Wolf through the other, you have to see two views of every single thing you think and say and do every single moment of your life, both when you're awake and asleep.

"You have to realize that everything you do is ultimately important on this planet, to yourself and to every other being. You live in a web of existence that connects the sun and moon and stars and planets and oceans and land and animals and stones and plants and even you humans. Just as your planet spins on its orbit, you must maintain your own cosmic vibration to extend your own existence within that whole. At the same time that your every vibe is crucial to all other existence, the reverse is just as obviously true. You depend on everything else for your existence too. This is why service is at the heart of all you're seeking here.

"At the same time," Coyote went on, "this whole place will be under water in a few thousand years and nothing you do today is going to matter a whit. Your work is worth zippo. Makes you stop and think, doesn't it?"

"Uh, well, I guess," I mumbled. I just couldn't quite get on even footing with Coyote. "Look, I think you're losing me. Let's go back a step. Why did you say that Comanche Jack and Wolf are liars?"

"Because they told you there was a proper way to live on Earth. They said that if you aligned your outer existence by channeling your perceptions through appropriate reflection, focused projection, and grounded balancing, your personal history would no longer bind you to behavior patterns that would limit your possibilities for dreaming, relating to others, and making appropriate choices. They said you would come to experience the inspiration, organization, transformation, and instinctual intellect that would illuminate your holographic access to the place of all phenomena. They told you this process would open you to the presence of constant synchronicity, heightened imagination, empowerment to pursue your dreams, and the freedom to live and die without the bondages created by your mind during every moment of perception. Am I right?"

"Sure. They told me that. But why do you say that's a lie?"

"Because, baby human, if all that's so true, what are you doing right now, right here? Why are you lying asleep in your body, physically helpless, emotionally disturbed, mentally at wit's end, and spiritually confused? If that pathway's so powerful, why do I have so much effect on you? After all, that pathway is what led you to me. The best test for any pathway is to take it to the limit, right? Well, I'm the limit. What kind of test results are you getting? Is Twenty Count helping you right now? If not, why not?"

Coyote had stunned me. I found no answers to the challenges. No weaknesses in the arguments. Perhaps this was the truth. Twenty Count was pointless. There was nothing to learn in it. After years of seeking and traveling and studying, I'd found the primordial flash and the message here was that I'd been wandering in a maze of futility.

"Are you despairing?" Coyote asked.

"Almost."

"Save your pity. You don't need it. Maybe you can stumble your way out of this maze as fortuitously as you wandered in. Maybe if we search hard enough, we can find an answer. Are you willing to try?"

"I don't think I have a choice."

"Great. I love volunteers. Now here's the scoop. You have to learn to lie, to deceive, to pretend, to manipulate, to get the upper hand on other people, and then lower the boom. You've got to take control of your daily life in just this manner."

"What are you talking about?" I asked. This was the weirdest advice I'd ever heard from a spirit guide. Maybe I'd tuned in to too many voices. Maybe I'd found one that I really didn't want to trust.

Coyote read my mind and countered my thoughts.

"Some people get one voice of guidance, others many. It doesn't matter. What does matter is developing the ability to test, trust, and utilize the full potential of this essential inner resource. This is not about channeling, nor does it concern the supernatural or even the unusual. It's about your ability to listen deeply to your personal inner languages. Your choices are whether or not to listen, how to interpret what you hear, and how to apply whatever you understand. This requires that your everyday life be molded into a world where this hearing both validates and is validated by direct experience. After all, both saints and murderers hear voices, and you need a way to ascertain the validity of what you perceive."

"And you say I've got to learn to lie in order to validate my experiences with the inner voices? It seems distorted to me. Maybe you'd better take this one step further. What's the bottom line idea behind all this?"

"That's easy. It's Twenty Count."

I tumbled into a space of complete mental silence. I could say nothing, ask nothing, think nothing. I just sat there. Waiting. Silently waiting. Time seemed to pass, but I wasn't sure. Coyote spoke again. Every word imprinted on my mind like a laser cutting marble.

"You have dreamed the ancient heart as utopia, but your imagination has not acted to create utopia. You have fantasized the ancient heart as the place of all answers, but your mind has not formulated penetrating questions. You have hoped the ancient heart to be a comfort zone of personal freedom, but you cling to the slavery of reliance on a systematic study of a system of numbers to force your inner voices into mystical revelations that circle back on one another. Your only chance to survive your fantasies and imaginings and hopes is to let go of your vision of the ancient heart. Accept that it is nonexistent. Accept the ancient heart as a nothing.

"By releasing every notion and expectation you've ever associated with the ancient heart, you'll find yourself reentering the everyday world once again. You'll awaken in your sleeping bag among the trees and flowers and mist and raccoons. But you will be different. You will have visited with me and talked with me. You will know that the next time you approach the ancient heart, you must remember who I am and how I speak. Then you will be able to speak effectively yourself. You will be able to hear, to see, to understand, to think more effectively, both out in your world and here in mine.

"Go home and sit down with your Aztec Sunstone. Meditate on it and ask yourself, 'Where did these ancient people discover the seeds of their cosmology? Why is it relevant to you today? Do all patterns of the past repeat endlessly in the present and the future? Or can you change them?'

"You don't want to be a liar. You cringe at my suggestions for manipulating and deceiving others. But this is what you have been doing every day of your life. You have been perpetuating old patterns of chronic beliefs that you inherited and have never challenged. You believe that you must apply mental interpretation to every impulse you receive from the universe, be it taste or sound or sight, through a process of reflection. This leads you to conclude that some things exist and others don't. Furthermore, the things that exist seem solid with absolute borders. If they're solid, you think, they're absolute, finite mass, not vibrating with life or light, and therefore they have limits. They're going to die, to vanish from existence without a trace. Because you perceive that you have a physical body, you think that you are just such a thing. You feel you're on the very edge of annihilation at every instant. This so terrifies you that you commit every ounce of your energy and imagination to creating an absolutely fascinating chain of avoidances and denials in the form of personality traits. You then focus all your communication with your outside world through this fantasy-based personality. Everything you say is based on illusion, avoidance and denial. You, my friend, are a liar."

My mind held silent as Coyote spoke. I absorbed the teachings but did not attach to them. They were clear and obvious rather than complex and obscure.

"So," Coyote went on, "if you're going to be a liar and manipulator and set up other people in your daily communication, you've got to learn how to do it right. That involves realizing that this outer world you're living in and dealing with is just simply not all of existence. In fact, it's a very small part. Humanity is a little smudge of life on a tiny little planet in the Milky Way that's just one self-organized entity in a cosmos that has existed for at least many billions of years and whose outer limits are completely beyond our imagination. And on the more immediate level, this will all be under water before very long at all.

"The important thing to understand at this moment is that as you deal with your outer world on a daily basis, you must continue to function through your personality, and everything you say and do will reflect in some way or other certain common beliefs that are observed to some degree by one or both of the people in communication. These beliefs are going to be based on the same fantasy of imminent annihilation that I just explained to you, and therefore your comments, no matter what they are, lack truth. Therefore, they will be lies.

"Furthermore, not understanding that you are lying when you speak with your inner baggage attached to your words also means that when you listen to others, you're then attaching your baggage to their words. So whatever you hear has untruth connected to it. In this way, you make liars out of those who attempt to speak truth to you.

"My message to you is that as you lie, just know it. See through your own words and the words of others. Accept them as meaningful only in the context of the moment, and keep your inner eyes and ears focused on the much bigger picture, the ancient heart. Then you can trust your inner voices. Then you can listen to people like Comanche Jack and guides like Wolf and hear their messages without attaching your inner expectations and denials and avoidances to them. And that way, you won't be making liars of them just by listening to them."

Abruptly, Coyote vanished. I could think again. I was flying comfortably astride Fox. I surfaced back within the primordial medicine wheel where this trip began. Fox had flown me home from the heart of the mystery itself. I had visited the actuality of Zero, the creative center, full of vibration and wind. I held to Fox and calmed myself. I inhaled and exhaled, realizing that I was passing back and forth through the opening of the ancient heart with each completed breath. Inside the heart, the flash existed. Outside, I did.

The ride was finished. Fox eased me into simpler sleep. Again, I sat cross-legged, knowing full well what had passed and what was to come. There was no one else there. I remembered the previous circle, though. I remembered it well. Many unique beings were sitting here with me. Each was a number. My number was Twenty.

It still is.

The new wait starts quietly, in the silence of no-thing. Perhaps I actually am nothing more than this silent wheel, a void around which to rally consciousness, a point in a center waiting to be. If I can realize myself as this point, I am unlimited as to how far I can travel in the next flash. I can create myself into a point that maintains its cool and calm focus, no matter what follows. I am stability amid a roaring turbulence of instability, silence amid chaos.

I sit, waiting for the flash. This time, it will gleam within me, not outside. I need chase nothing. The opening will reveal the deeper essence of magic mind, the pulsating process of going through the creative center of the ancient heart. This is like waiting for death, a peaceful, exhilarating anticipation. Faithful Fox lies vigilant in me too, ready to fly ahead on the next daring journey. I sit, waiting. I now know how to listen and speak. No one else is in my circle.

This time I am alone.

LUNAR-SOLAR INSIGHT

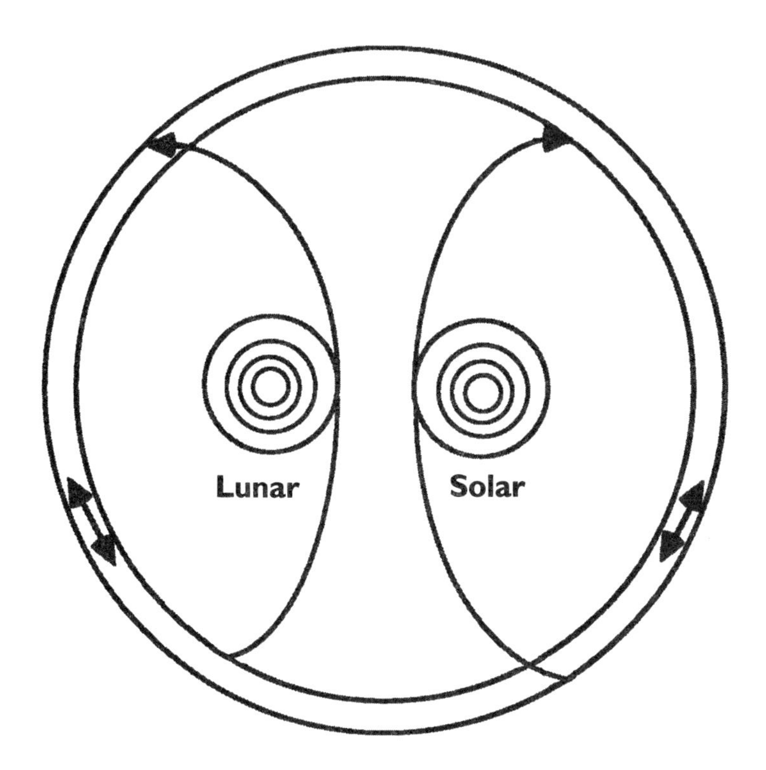

"...Though mountains danced before them they said that G-d was dead. Though his shrouds were hoisted the naked G-d did live. This I mean to whisper to my mind. This I mean to laugh with in my mind. This I mean my mind to serve till service is but Magic moving through the world, and mind itself is Magic coursing through the flesh, and flesh itself is Magic dancing on a clock, and time itself the Magic Length of G-d."

— Leonard Cohen, "Magic is Alive" from *Beautiful Losers*

In Finite Chaos

All stimuli suspended, I floated naked on my back in salt water solution. Total darkness, no sound. I thought of being there, in the sensory deprivation tank. No one else around at 10 a.m., Christmas morning, three hours before I was due at the holiday feast. An hour in the tank would refresh and relax me.

Time stopped and Christmas vanished. The water no longer imposed sensation on my skin, and I ceased to require support. I did not sleep. Or perhaps I did. Delving into perceptual nooks and conceptual crannies previously unexplored, I sought disembodied filaments that separated from light rays and lodged in my inner mind during the singular light of the primordial flash.

Deep dark blue was mind. Struggling, either an instant or forever, vision accepted the blue. Another sensation registered, but through hearing. Inhaling initiated sound vibration, exhaling completed it. The two-step breath impelled perception from one side of the ancient heart to the other and back again, aligning vibration into sound. Then the music came.

My response to inner music was more profound than to any vision. Music's sustenance of its own existence through sound reverberation and resonance produced a purely pleasurable sensation. I also identified music as one with breath. As hearing intensified into innovative forms and patterns on this Christmas day, my appreciation grew. I reflected that the gift of vibrational recognition within octave-based guidelines is the first and most powerful shared activity available to all past, present, and future listeners. Music initiates sharing and necessitates other beings with whom to share. To love music is to love giving. I floated in heartfulness.

"You want to tune into other realms, man, you better make your own music now."

The voice burst into a roar of laughter like a blustery swirl of wind in my ears. A smoky vision formed into a blue image of a man. He wore a beat-up leather hat and wire-rimmed spectacles with reflector lenses. His hair was braided onto one shoulder. He was a hippie and looked a lot like Comanche Jack. Talked like him too.

"Hello there, brother. My name is Holygram, the spirit of Fifteen. Welcome to my pad here in kaleidoscopia. Welcome, man, look around and listen about. It's a wild world in here. Anything you can imagine, anything Creator can create, it's all here. See the thoughts. Hear the shapes. Feel the colors. If you're looking for secrets, you've dreamed to the right place. Now all you have to do is figure out what's secret and what's not. Otherwise you'll never find your balance."

Half-stunned and half-thrilled, I asked, "How in the world can I do that?"

"In your world, it's difficult, maybe impossible," Holygram

said. "Your world must grow. You need a more powerful resource than you're using. Right now, even with your music, all you're doing is taking in and processing sensations. You need to put out. Develop your reason, your mind. Think, man, think. Be cool."

"Hey, I'm cool. I'm cool," I protested. My psychic composure was returning after the shock of his appearance. So this was one of the twenty guides Wolf had told me about. Well, he was a colorful one to start with. "What's the story in here, Holygram?"

"To get the story, you gotta see the world through these lenses," he laughed, pointing to his silvered shades. "Everything's visible, even the music. Especially the music. But you couldn't take it now. Maybe another time."

"No way, brother, I'm ready now. Let me try them. Just for a minute."

"If you say so, man. It's up to you."

My dream hands took the sunglasses from Holygram and fitted them over my eyes. The blue of the tank trip instantly swirled into twin spinning wheels encompassed by a larger wheel. The music turned discordant and grated on my nerves. Energy was flowing in all directions.

"Looks like chaos, right?" Holygram asked.

Without being told, I identified solar energies on the right and lunar energies on the left whirling in opposite directions, a rhythmic circuit evolving eternally into its source. Two synchronous forces unaware of their separateness or differences. Suddenly the primordial flash itself invaded like lightning.

Time and space instantly aligned the finite order into existence as two halves. The solar flash shone on the lunar body within their shared field of consciousness. The solar was male, the lunar female. Their act of joint recognition of a magnetic draw to reunite was sexuality. This awareness aligned male subject apart from female object, relegating memory of the original synchronous deep into a dream state that would endure so long as the chronology of time and space dominated perception.

"See, they both understand that order–their vaguely recalled image of original perfection–exists only as dream now," Holygram explained. "So the only thing that makes sense to both is to try to return to the dream while maintaining the spinning flash, the worldly existence that provides the opportunity for that return. One indelible memory from the flash is unity, the flesh of two as one. Thus, reunion's got to be an essential element of the return, an element of magic sought either actively or passively through the movement of flesh. Like they say, the world will always welcome lovers, whatever time goes by.

"Now it's time for you to meet chaos and order. Chaos pops up as a landmark on fantasy road when the perception of human disorder directly contradicts the dream of order. The struggle to recapture the lost order must somehow transcend this chaos. It's not easy, because chaos–which is above us, below us, in front of us, behind us, to the left of us, to the right of us, and within us–denies the reality of the dream order. That order, chaos proclaims, exists only in the hazy, escapist, Pollyanna world of the unrealistic and unreachable realms. Such an order is not alive. If humanity perceives that highest conceivable order as God, chaos echoes Nietzsche by proclaiming that God, indeed, is dead. Chaos got a lot of publicity about this in the 1960s.

"Stay with me, OK?" Holygram said as I adjusted my visionary lenses. "Male is open to this belief. After all, he's just watching female as object. True, he feels the need to move his flesh in union with female flesh, but he thinks that's just some inexplicable thing called man's nature. No need to make such an intense deal about sex, right? It's just an urge unconnected to God or anything else. And besides, God's probably dead. Said so on TV. Screw it.

"Female realizes chaos is a liar. She's not primarily a watcher, so TV makes less of an impression on her. Embodying the physical apparatus to reproduce life—to become again—female remembers original order more strongly than male. Sexuality, flesh reflected in flesh, is an intimate memory of the flash for her, a replication of the creation process within the matrix of the ancient heart. Life itself is the movement directly from and back to that matrix."

Holygram had turned into quite a lecturer, his monologue supplementing my visioning perfectly. I realized that my vision had activated a different voice/hearing connection between the two of us than had existed before I put on the glasses. For a split second I could see and hear Coyote saying, "See, you've converted Holygram from a liar to a truth-teller. Keep watching. Keep hearing."

Holygram kept talking. "To extend existence in order to facilitate her ultimate return, female initiates an active passivity by dancing a molecular dance of attraction for the benefit of male. Her alluring vibration of hips and lips summons up a dream state that enslaves male, who no longer believes in and, therefore, is defenseless against the dream. Male, now a willing disciple of chaos, copulates with female. They come in celebration, despite their different agendas. Female wants to become again. Male merely wants to come again. At whatever point they realize their purposes conflict, they simply subvert their individual priorities in order to continue the path of immediate allurement, then copulation. Though subverted, the priorities of chaos and order do not disappear. They transform into hidden agendas. Still, male and female have conducted their cosmic symphony, hummed their cosmic melodies, and danced their cosmic dance.

"Remember your Twenty Count numbers? Male is the role model for One. He starts out as observer and then merges into fantasy-prone allurement and reproduction. He watches the cosmos explode into an ocean of new beings, some male and some female. There's no sanctity in copulation and there're lots of fish in the sea. Chaos is turning out to be a most enjoyable guru. Much cooler than the late God, for sure.

"Each new being inherits a spark of the dream of returning to the matrix as well as the unsettling suggestiveness of chaos. Each perceives a personal universe based on either reproducing or having a good time. Depending on personal predilection, each aligns with one or the other hidden agenda. Male or female, all beings are driven either to get it right or to just get it."

Holygram was finished, and he let me know with a laughing roar that shook his shiny specs off my eyes. His lecture remained in my memory banks as my eyes readjusted to deep dark blue mind. In a little while Wolf appeared.

"You realize, don't you," Wolf counseled, "that you saw all you saw and heard all you heard today because you weren't obsessing on either order or chaos. You were open because no stimuli were prodding you toward any particular concern. Then, and only then, did Holygram reveal himself. It's good to realize how these things happen, because the vast majority of human beings live with one or the other or even both constant obsessions. Their parents were that way, and so are they. It's difficult to escape the hidden agendas."

"Just one question," I said. "Are you and Holygram saying that all human beings, everyone on Earth, are trapped in this conflict and these hidden agendas?

"Interestingly enough," Wolf answered, "there are some people who live free of both obsessions. Rejecting chaos, they open themselves to more possibilities because they sense the full power of the ancient heart. Ignoring the compulsion to identify with established order, they look out on existence through an alternative viewpoint. Some call it the third eye.

"Trust and openness in communication guide these strange children who pass through the same greedy and demanding environment as do those committed to order and chaos. They acknowledge the powerful force emitting from the ancient heart as will. In turn, they too possess will. They may do things that seem drastic, like leave their homes and forsake their traditions. But they love easily and acknowledge a gentle god or goddess who frees them from the compulsive need to escape this lovely world by avoiding direct perception.

"Still, these unusual folks' reality is no easy stroll. They're a minority no matter where they travel in this world, and few of them are successful in business. Their nervous system shakes a little differently. There is, however, a powerful inherent potential: they want to be of service.

"These weirdoes are the embracers. They love to hug. You knew one very well back in Santa Fe, if you could only remember. Now wake up. Wake up. Time to rise up out of the water. Just stay away from those hidden agendas."

Volition jolted my body like an electric shock and I splashed out of the sensory deprivation tank. I had never felt lighter and was totally alert. My entire being vibrated and tingled as I showered, and I marveled at the clarity of both revelation and remembrance. I had met Holygram, viewed the cosmic dance of sexuality, and heard the cosmic music. What a morning.

Given that I had entered the tank about 10 a.m. and the entire adventure seemed merely an instantaneous visit to kaleidoscopia, I couldn't believe how starved I felt until I noticed the pitch black sky outside the window.

The wall clock showed 6 p.m., and Christmas dinner had been gobbled up long ago without me. My tank trip had lasted eight hours.

COYOTE ROAD

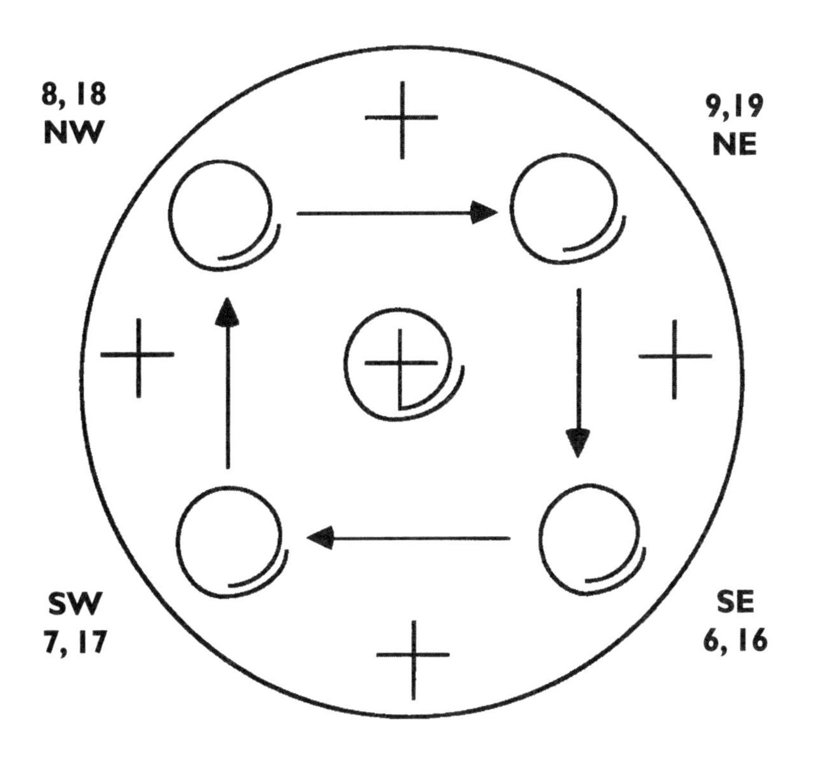

The cross
with no circle
creates pain
with no power.

Embracers

Dusk enveloped the foothills northeast of Santa Fe. Sunset orange and pink and purple lowered over the horizon. The woman with long black hair sang an olden chant over the abalone shell of burning sage, the smoke twisting like a mystical river toward the heavens. She finished her hymn to the dying sun, blessing its journey into darkness and praying for its resurrection tomorrow. The pattern of eons meant nothing to her and she took no chance. Life's moment was fresh with the sense of what brought the morning light. Embracers feel such things.

Sheila Dawn lived in an abandoned trailer on scrubby brushland about an hour's drive from Santa Fe. A couple of other shabby trailers, a half-dozen dusty cars, and a few huts hovered in relative proximity in a quiet land where electricity did not intrude. Someone in Philadelphia owned the land but never visited. Sweet people here, gentle and giving, reclusive and troubled. Victims of themselves and their parents' seeds and vines and knots, some drink and some do drugs and some meditate. They participate fully in the mystery game of the ancient heart, and they play, without much conviction, the identity game. Wisdom is their object of desire, but it seldom manifests for them because participation in the world demands too much sacrifice. They lack means to achieve their ends.

Embracers, beings of light and wind, live everywhere, indifferent to both order and chaos. This quality alone makes them versatile, flexible, sometimes enduring. It does not necessarily make them effective, because it also leaves them vulnerable. They too must search their way into the ancient heart, but on a different path from fantasy road, which they abandon as just too painful. They reject establishment values in favor of more heartfelt standards.

Embracers also live in condominiums and town houses. Some are professional protesters who travel the radiation path of the nation's nuclear facilities and live through the support system of an informal underground. Others program computers. Some work nine to five, others beg on filthy streets and pray in candlelight. Some are vegetarians, others eat salty meat. Their outer beings offer no classification; you must look inward. Like all other measurable beings from triangles to chromosomes, embracers exist in varying forms, sometimes inhabiting the same bodies as chaos/order people. Possibly you can see one in yourself. At least I did, and I left journalism.

While fantasy road is based on the cross-directional pattern of east, west, south, north, coyote road goes southeast, southwest, northwest, and northeast, a clockwise square within a circle of gentle spirals emitting only mild force. This alternative lifestyle focuses on letting go, living dreams, releasing relationships, and going with the circular flow. The ninety-degree angles of the squared path offer both turning points and blockages. The agendas of the hero's will and the perfect servant thrive like truck stops along coyote road.

Southeast, where the bondages of history evaporate from Six to Sixteen through the process of incorporating synchronicity with chronology, is the place of letting go. To realize that one's history need not determine every aspect of one's being is to transcend limiting forms. Letting go can apply to ideas, families, jobs, whatever.

Moving to southwest, where dreams access the place of all phenomena at Seven and Seventeen, embracers seek guidance to replace the binding force of history. Imagination opens up that which tradition condensed tightly. Distinguishing between living one's dreams and living in one's dreams is the trick. The place of all phenomena is a great place to visit, but you don't want to live there.

Embracers face their karmic ties in northwest at Eight and Eighteen, and attempt to release them. These folks are serious about not encumbering other beings. It's just that karmic invisibility is not readily apparent to the untrained seer. Although embracers may not seek control, for instance, many have yet to let go of wanting approval, so they still tie themselves into the mentality of others. Balance is difficult to maintain, and can be indistinguishable from indifference. It's simply tough to unknot those ties that blind.

Still, embracers prefer to live in bliss, releasing original guilt and shame from the ties that blind and opening themselves up to the more expansive possibilities of inspiration and imagination on their continuing exploration of coyote road. This leads to northeast, Nine and Nineteen, where the choice is repetition or an inward-turned spiral. Instead of subscribing to the hidden agenda choice of chaos/order, embracers prefer the wind and vibration of light and movement. They seek the will of the ancient heart by floating with the flow, ignoring alignment and order. Their arms are wide open and compassion is in bloom. They challenge perception, ignore reflection, undermine projection, and reject balance. Their tendency is to open themselves to whatever will be, to allow the winds to carry them deeper into the multidimensional wheels of life.

Coyote road feels far more adventuresome to embracers than fantasy road. Still, every pathway offers the possibility of both reality and illusion. The potential of embracers lies in finding their calling as healers and service providers to others, but their greatest dangers await there too. Coyote road, though a path apart from fantasy road and its pit stops of chaos and order, leads embracers into the hidden agendas of hero's will and perfect servant. Though these agendas bypass sexual roles, they are still enticingly attractive to embracers vibing in bliss.

The agenda of the hero's will involves a deep desire to control. This manifests as an individual wanting to actively dominate suffering and change the world, rather than remaining overtly passive as a perceiver who observes a painful situation working itself out according to some slower, more involved plan that may be so imperceivably subtle as to exist only as a glint of inner knowing. This agenda stems from a blind faith in one's power to change the world, being convinced that "I am the chosen one."

This is a particular trap set by Coyote for the seeker tripping out on the bliss of a recent religious awakening. One of the most common experiences for seekers entering the religious ring is to receive messages from the great beyond. Psychics do it, preachers do it, dreamers do it, anybody can do it. This hearing is so profound that the seeker forgets about seeing. As the recently inducted seeker flushes in the excitement of receiving information, an act that any television antennae can perform, Coyote sends a message in reverent tones worthy of mass media fund raising. It usually goes like this:

Coyote: "Seeker, you are the one."

Seeker: "What?" Indeed, what on Earth?

Coyote: "I said, you are the one."

Seeker: "I'm the one?" Religious reference points begin to stir.

Coyote: "You are the one."

Seeker: "Wait a minute. Are you saying what I think you're saying?" My God, am I the One?

Coyote: "You are the one."

Seeker: "Are you telling me I'm the One everybody's been waiting for? Am I the new Messiah? Am I the reincarnated Savior? Am I that One?"

Coyote: "You are the one."

Seeker: "I can't believe it." Well, yes, I can. I've always had a feeling about myself. And now this proves it. How can I argue with this voice from beyond? So I suppose I should tell other people about this, yeah? And then they'll want to follow Me? Is that what I should do?

Coyote: "You are the one."

Seeker: "All right, all right, all right. I get it now. Thank you, thank you, thank you. I'm going to get that church started now. And these people are going to just love Me. And God, do you think I could make some money along the way too? Would that be all right?"

COYOTE: "You are the one." Coyote also murmurs something else, but Seeker is too busy setting up an accounting system to catch the nearly silent message.

SEEKER: "Yes, yes, I am the One. And my congregation will love Me and obey Me. That's the way it should be. We're going to be one big, happy family. We'll all worship together just like I worship now, and they'll all honor and obey Me. The very fact that I've heard the voice of God proves that My way is right. After all, it got Me this far."

The hero's awareness heard, but did not see. Seeing would have revealed Coyote grinning in the shadows, and also those final soft words of Coyote's message:

"Yep, you are the one. And I am the one. We are all the one. We are all primordial awareness. That's it. That's all I meant. You have to use my reference points, not yours. It's that simple. You thought you understood me, and now you're self-deluded."

The existence of this trap in the hidden agenda of the hero's will is no fantasy. Its reality has been proven dramatically by tragic figures like Jim Jones and David Koresh. But beings in bliss don't seem to get it. Instead, they willingly affirm the illusion by following the self-deluded hero until they become its victims.

In the hidden agenda of the perfect servant, a deep need for approval from something or someone of great power inspires the seeker to bow to a higher will and enter into the syndrome of putting others first. This allows the individual to act out the image of unworthiness through service. This is opposite to the hero's will agenda as it embodies a belief such as "I am nothing before the awesome glory and power of God;" or perhaps "I am helpless amid all this suffering on Earth;" or perhaps "I am incapable of making it in the dog-eat-dog world of private enterprise and competitive endeavor, and therefore I perform the most humbling service I can dream of, like feeding the beggars who are manifesting an even more severe aspect of the perfect servant agenda, that is, being the enablers who require perfect servants in order to survive."

The danger in the hidden agenda of the perfect servant is that the servant progressively gives away energetic essence while the enabling receiver of service slurps up that energy like a vampire. It's a never-ending cycle. This agenda manifests in many a family, and family members are extremely generous in carrying the pattern over into virtually all their life activities. Thus, organizations such as churches and communes and nonprofit service groups are teeming hotbeds of hidden agendas of the perfect servant as well as breeding grounds for their vampires.

The hidden agendas of chaos and order, remember, exist as underlying motivation for male and female beings trapped on fantasy road, replaying over and over the melodrama of separation and return. But just because two agendas seem diametrically opposed does not mean they represent different points of view. One agenda honors chaos and another honors order, but both obsess on honoring glimpsed flashes of split-second perceptions on fantasy road. In the same manner, one embracers' agenda may stress serving others, while another focuses on aggrandizing separated self, but both feed on the drive of mission-zeal purpose and both draw inspiration from within. Both are driven by ungrounded bliss, that is, bliss that is uneducated, inexperienced, naive, self-centered, and imbued of great fear.

Bliss, in order to be a useful source of power, must be grounded. On the map of hidden secrets, for example, the cross-directional rays creating four directions must be anchored within the wheel. Otherwise, all the rays are incomplete, their shine dull and non-illuminating. The medicine wheel is complete only when the circle and cross exist together, just as any entity is complete only when it exists in its own environment. A human being without oxygen to breathe or without protective shelter or without food for sustenance ceases to be a human being. The cross without the circle ceases to be a wheel. No wonder that in Western religions, the cross with no circle has become an emblem of powerless pain, an extension of the hidden agenda of the perfect servant.

Like most embracers, Sheila Dawn mixed agendas. Her embracing was incomplete and she constantly battled her learned need for order by creating chaos. She burned sacred plants and chanted to the sun to add her own will to the universal power to bring the fire back tomorrow morning and forestall the chaos that created such need in the world. She was extremely kind and she made it so easy to accept her kindness.

"I have been looking for a book for a long time," she said as the sun went down. "I don't know who wrote it and I don't know the title. Can you help me find it?"

"I'm not sure," I said, breathing in the pungent tang of sage. "What's it about?"

"Old Ghost," she laughed. Her dark eyes reflected the night's shining Venus. "Have you ever heard of him? I'm just asking because you seem to like books. I thought maybe you'd run across this one."

"I don't think I know the one you're talking about. Can you tell me a little more? When did you see this book?"

"Twenty years ago when I was a little girl. It was a story about Old Ghost. He was a spirit who appeared and talked to people, but only one little boy could hear him. People thought the boy was crazy because he'd suddenly start listening or talking to the thin air. But Old Ghost told him great things, secret things. The boy learned so much from him and then grew up knowing all this wonderful stuff. All about life and how to think and see. I was in grade school when I read the book and I loved it, but now I can't remember any more than that. I can't remember what Old Ghost told the boy, what his secrets were. I've been trying to find that book for years. I go into libraries and used bookstores and walk up and down the aisles, searching the shelves. It makes me kind of crazy sometimes, but I need to find that book. Anyway, I guess you don't know about it."

Though I'd never heard of Old Ghost or his story, I knew his secrets were the wheels spinning in Twenty Count, the aspects of light hidden in the folds of mind. Some forgotten inspiration lay dormant in both Sheila's and my memories, something with different details but identical essence. I wanted to remember as much as she did. I promised to look for the book in California.

"So you're really going to California?" she asked. "Why? What is there that's not here? The sunsets can't be more beautiful."

"There's an ocean," I said. For a long time I reflected, trying to match words to reasons. Wolf stood at my shoulder, a careful and purposeful guide. The woman was looking for something I understood but could not quite remember. I was tracking my own way into mystery. I told her a feeble, evasive truth. "Some secrets live in California and nowhere else. I think my own Old Ghost wants me to go."

She said she understood. The sage burned to ash as the last light faded into night. We watched the distant hills a long time.

Early next morning, I drove to Sandia Peak and rode the ski lift to the snow-capped summit. I hiked a short, steep trail where ice melted and wisps of mountain sage peeked through snow. The clouds parted and the sun appeared. Sheila Dawn had given me a gift called Old Ghost, we had discovered a common memory, and I had learned what brings the morning light.

"The secret," whispered an inner voice as I turned away from life with a gentle embracer, "is the art of passionate detachment. It's by far the most potent secret you've encountered in the game so far. In fact, it's a matter of life and death. Your own."

REACHING OUT

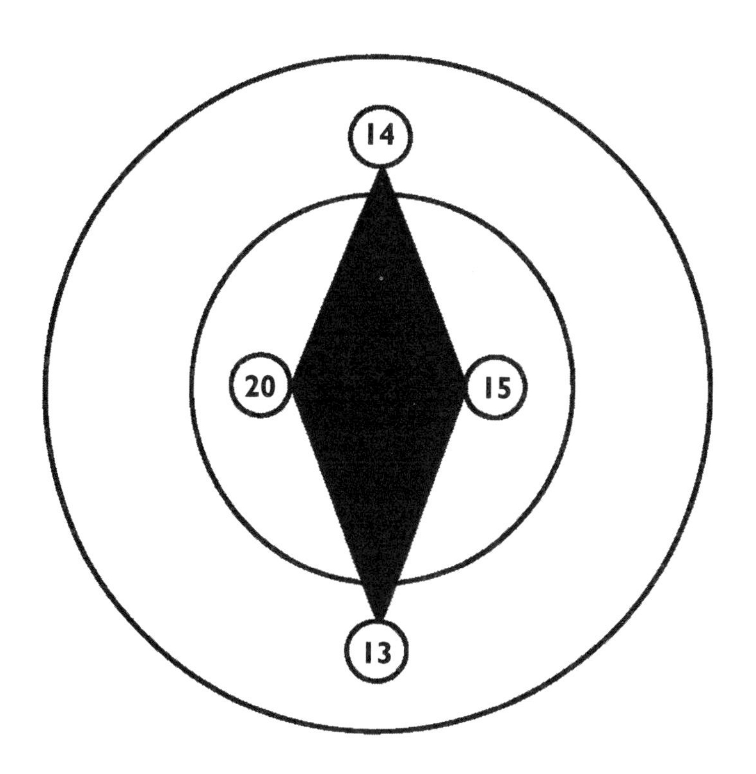

As seers
we stand.
As listeners
we open.

The Art of Old Ghost

I began wandering used bookstores from New Mexico to California. Just who or what was Old Ghost? All Sheila Dawn remembered was that someone once wrote about him. This, somehow, seemed the one question above all others to resolve. In a single day, my obsession with finding Old Ghost overwhelmed my desire for clarity. The inner quest assumed an outer focus.

Then one night on a hillside, Wolf came to the rescue. I was lying awake in my sleeping bag, staring at midnight stars and listening to distant howling. "I want to introduce someone," Wolf said simply. "This is Old Ghost, the spirit of the ancient heart."

It was so simple and it made sense. Old Ghost was a spirit, a guide from within Twenty Count–and the guide of Twenty, at that. I remembered to look as well as listen, and the night sky's ebony sheen heightened the shimmery outline of Old Ghost. Long hair streamed down over the shoulders, the smiling face aged in beauty and strength. Old Ghost may have been man, may have been woman, may have been androgynous. It didn't matter at all.

"I thought you'd come last," I said. "I thought I'd meet nineteen other guides before you."

"You're possessed by the evil spirit of chronology," Old Ghost laughed in a soft but pervasive voice. "Think about it. If I came last, how could I guide you back to myself? You'd spend a lifetime wandering around Twenty Count without ever knowing where you were going."

"But you didn't come first either. Why did Holygram get here before you?"

"You're still stuck in chronology. Holygram appeared first so you could see the big picture and appreciate how much you need me. I thought you'd like an overview of this operation. After all, you've got a vested interest."

"How can the spirit of the ancient heart speak to me in such up-to-date English?" My suspicious side would keep testing visions and voices forever.

"Chronology again. You're really obsessed. It's all happening at once, little friend. If you accept that Wolf speaks directly to you, how is it you question that the spirit of Twenty would not be familiar with your culture's current linguistics? After all, I represent your return, right? Therefore, I'm your future as well as your past. Gives a new meaning to ancient, doesn't it? I am Old Ghost, and I am your future. Now, let's proceed with that overview."

Before my eyes, which may have been either open or closed, a vertical black diamond appeared within a circle.

"This is your existence in the abstract realm, the world you have chosen to live in. You selected this realm when you committed to your quest," Old Ghost explained. "The base point of your diamond is Thirteen, baby death. The top point is Fourteen, instinctual intellect. That's just another way of talking about smoke knowledge. It's reaching the point where your body knows how best to guide your life, and mind goes along with it. You become smart like Wolf.

"This is why, in tribal societies, young people are counseled to learn life by adopting and studying an animal guide in the wilds. Animals simply walk the straight line. Their instinct involves being in touch with Mother Earth. They walk on the ground and they know it. They do not have to think about it. They're grounded into direct contact with the spirit world through their awareness, their ability to move and choose and follow their inner guidance without a blocking thought.

"You, however, as a human, possess an additional aspect to your nature. You also have the capacity to reach out with your mentality. You can use your perceptive and reflective capacities to avoid falling prey to the hidden agendas along fantasy road. You start this by reaching your left hand out to your side and taking hold of Fifteen, Holygram, and then reaching your right hand out to grasp Twenty. That's me, Old Ghost. This reaching out creates the vertical diamond of potential of the abstract realm. The blackness within the diamond is the unknown abstraction within you. This is the territory you must explore after baby death.

"It is at this point that you begin to perceive the words of the guides. You begin to hear. As a seer, you stand upright. As a listener, extending perception to hear the messages of existence as they pass back and forth through the creative center, you open yourself to the spiral that returns continuously to Zero. There you can commence to examine source and the reality of existence."

Old Ghost paused. Every word had imprinted as I listened in the clarity of night. I wanted to ask questions, but as one after another formed, Wolf whispered the same assurance repeatedly: "Let go of chronology and all will be clear."

"That's true," Old Ghost confirmed. "When you release the binding vines of chronological time and the limiting boundaries of what you perceive as absolute space, questions dissolve into the answers of silent instinct. This also is smoke knowledge, as I explained. Before one can enter this realm, baby self must die, maybe once, maybe a thousand times. But there will come one particular baby death, and after that one, you will begin to hear. This happened back on Algonquin Peak when you surrendered your past. It's strange, but when you surrender like that, it doesn't really matter whether you know what you're doing or not. Actually, if you knew what you were doing, you probably wouldn't do it. But that's what Red Thunder Cloud taught. Just say it. Just do it. By the way, Red Thunder Cloud preceded Holygram as a Twenty Count guide."

This revelation was startling. Pressure suddenly built up in my skull, breath came with difficulty, and my body submitted to a strange paralysis. Years of fears pulsated throughout. Chronology demanded an answer to the question that forced itself from my tongue.

"That can't be," I protested. "I met Red Thunder Cloud before I even knew about the twenty spirits. And besides, he's a living, breathing man, not like you, and not like Holygram. He's outside me, not inside. Why are you telling me this?"

"Such limitations," Old Ghost smiled. "Such beliefs. Your worlds of flesh and spirit are so mutually exclusive. Relax. Let a little light into your obsessive mass. You need to cultivate your art of passionate detachment. Learn to laugh, at least a little, at all these things you believe so vehemently. Otherwise, you're going to have a difficult time getting to know all the other guides. You're going to end up possessed rather than blessed. All these voices will be rattling around in your mind and you'll wonder where they're coming from. And because you can't laugh at all the stuff you're holding onto as standards and self-image, you're going to experience all kind of paranormal turbulence. You'll think you've swallowed a red-eyed demon when all you're seeing is Coyote blinking back the reflection of your own perceived limitations. That's evil possession, and that's the devil in the mind."

I digested this admonition. The only remaining sensation was a chilling, long-lasting shiver up and down my spine. The instant I recalled Wolf's prediction that I would experience the twenty spirits in a shiver, my visitor continued.

"Learn to develop the art of passionate detachment," Old Ghost said. "When one isolates information and introduces it into the fantasy world, it doesn't translate directly into knowledge. It merely becomes more fuel for fantasy. You must master the act of detaching yourself from fantasy road, yet you must retain your drive, your zeal for life. It's tricky business. Detachment and passion seem like opposites, like the fantasy of projection that requires expanded awareness to understand that apparent opposites are actually complements to one another."

"What exactly am I detaching from?" I asked.

"Fantasy, of course. Your original reflected projection that you balanced into a version of reality. It's the mythology that you built your life on. And it's difficult to detach from it because it's so damned sticky. Every surface has this gooey stuff that binds itself to anything or anyone touching it. Quite an effective trap."

"Then how does one detach?"

"Passionate detachment is a matter of life and death, and it requires absolute ruthlessness. I whispered that in your ear once before, just as you were leaving your embracer friend. You have to be willing to break your own heart, turn your back on those closest to you, betray every trust, give up everything that matters. Sound familiar?"

"Sounds like baby death."

"Exactly. And it is a real death. In reality, you need perform no physical action nor take any particular stance that others would identify as an obvious change. But you must do whatever is necessary to liberate your inner receptivity in order to experience freely the most direct possible sensation of the original source. The being that emerges within you after baby death is an embracer. This entire process is all about empowering embracers, both within yourself and in the world around you. The most accurate possible interaction with the ancient heart heals both individual and community.

"To go through baby death is to unglue your mythology, that story of all your successes and failures during pursuit of mystery. There's a lot of sticky stuff to undo. But if that mythology can be released, that sticky stuff loses its hold. When it does, then the embracer within gets free, and it becomes your responsibility to care for that embracer. To educate it and equip it with power and tools to deal with the world."

"And Red Thunder Cloud was my guide for baby death?"

"Yes. His presence provided you an image to reflect on and project into your own life. A role model, more or less. It doesn't matter that you were a mature man when you met him. Children exist at all ages, and empowerment is as significant at ninety as at nine. If you don't believe Red Thunder Cloud was a Twenty Count guide, just try speaking to him when I leave. You'll find it doesn't matter if he's in body or spirit at this time. You'll still be able to communicate with him."

Old Ghost, of course, was right. Red Thunder Cloud, who still lives on this planet, speaks to me through dreams and meditation as vividly as any other of the twenty spirits. So I was not so surprised when I learned that Keetoowah, no longer on this planet, was my spirit of Fourteen, the place of silent instinct. The old Cherokee who taught the no-thing of the flashing light has stayed with me. He also appears more at peace in spirit than on Earth, less quarrelsome now that his pain has been assuaged. The two Native American healers anchor the opposite poles of my north-south axis.

Now the opening through which other spirits would appear became obvious. Contact had been happening all along without my volition or awareness. I realized late one night during a meditation that these voices had been present for years during meditations, dreams, and just about all other times as well. This realization came about amid a haze of words and images flowing in from somewhere other than my own thinking processes. I was not sure of the sources of the information, but neither could I deny the validity of the insight. I was certainly not afraid in any way, and the voices passed on information that proved to be accurate and valuable. Little by little, I learned to trust what I received.

Reflecting on this process of reception, I remembered this precise communication had been occurring as long ago as my memory could stretch, way back into childhood. Perhaps since I first learned to talk or listen. I just had never before recalled the process during the moments immediately following reception, or perhaps my mind accepted it as so innately common and normal that the process never stood out. My voices of guidance had always been there. For some reason I'd glossed over them continuously, moment after moment, for years.

This realization amazed and pleased me perhaps more than anything else in my life to that time. I stood and walked and laughed aloud. I suddenly felt starved, and during just that single millisecond's sensation of hunger, the entire realization nearly vanished. In the midst of creating a peanut butter and banana sandwich, I somehow recalled the amazing remembrance that had so inspired me moments earlier, and I also recalled that I'd forgotten it previously and had nearly done so again. Perhaps I'd actually remembered and forgotten it a thousand times–or more, perhaps four hundred million times. I dropped the sandwich and ran for pen and paper to record the realization and solidify it as part of my daily awareness. There are moments I still forget.

Remembering the presence of guidance reveals the significance of transcending the limitations of prescribed norms, which act as buffers against our deep memory. Breaking away from these norms is a matter of some difficulty and trickiness. But it can be done, and again the spiral offers an opening and insight not available elsewhere. Recovering deep memory is possibly the most critical step to success on the quest toward Twenty.

Deep memory refers to the secrets buried in the folds of wind and vibration in magic mind. As mind vibrates, it aligns these secret entities into clusters–material objects and mental concepts that we humans comprehend and name. This was the role Adam played back in the Garden of Eden, seeing and naming. We have been running the same inventory process ever since, but we still don't get it right.

Rather than remembering the direct essence we perceive, we interpret that essence into forms and actions that conform to our learned belief systems. As a result, we of the modern world have ingested so many interpretations and fantasies that the secrets are buried even deeper than ever before. There is space, so much space between our thoughts and actions, and that space is loaded with patterns atop beliefs atop fantasies atop patterns atop beliefs atop fantasies.

"Really," said Old Ghost that night under the stars, "it's just a lot of B.S. Get it? Belief System–B.S. And you wonder where you get your sense of humor. But as I said before, it remains a matter of life and death for you to master this art of passionate detachment. It is urgent to do it now. You've asked about detachment, but not the passion. Do you understand yourself enough to know why?"

"I don't know that I can answer you," I said. "But I think I feel the passion anyway. I don't think it's going to disappear. Maybe you can tell me why that is."

"Because this is the right time and this is the right place," Old Ghost said. "Because this is where the mythology is coming apart. And when belief systems die, the spirits with the most passion will shine brightest."

DEVELOPING PRESENCE

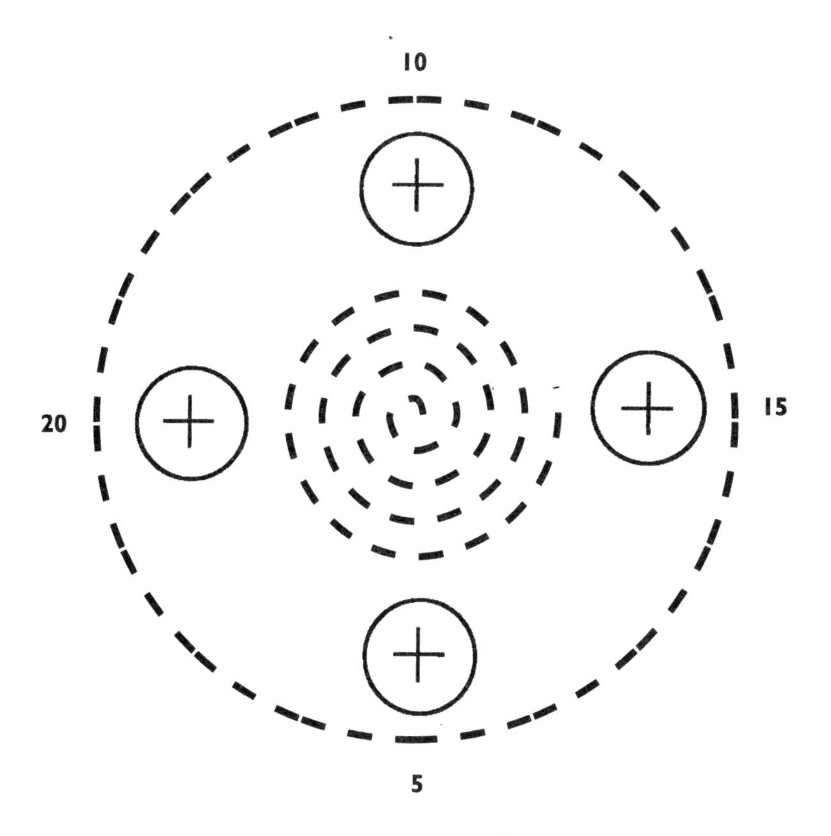

5 — Separated Self

10 — Expanded Awareness

15 — Kaleidoscopia

20 — Ancient Heart

Machinery for Change

It's coming to America first,
the cradle of the best and the worst,
it's here they got the range
and the machinery for change,
it's here they've got the spiritual thirst.
– Leonard Cohen, "Democracy"

Barely a breeze stirred that sparkling September morn in 1982, a few months before Algonquin Peak and my move to California. I was wolfing down a quick breakfast before heading for the Great Masters World Peace Conference at the Providence Zen Center when a shattering crash from upstairs resounded throughout the house. I went up the stairs three at a time.

There, on my office floor, lay the fractured remains of the foot-high plaster Buddha that had been resting peacefully for months in the window nook of the slanted roof. The window was open, just as it had been all summer and fall, but it would have taken a mighty gust of wind to blast the Buddha off his perch. There had been none. The statue had been my prize meditation piece for several years, and I was more than a little attached to it. My insides were shaking to the point of nausea as I swept up the fragmented body. But there was little time to reflect on my shattered companion. I was reporting on the conference for a magazine.

The Providence Zen Center crowns a grassy knoll near Cumberland, Rhode Island, ten miles north of Providence and forty miles south of Walden Pond. Spiritual masters from around the world were descending on the freshly painted complex and lake in this pristine, rustic setting. On this day of the shattered Buddha, I would meet and hear men and women whose communication capacities far outsped the limitations of my note-taking and reporting.

The occasion for this first-ever Buddhist peace rally in the United States was the tenth anniversary of Korean Buddhism in this country. The Providence Zen Center had been founded in 1972 by today's host, Zen Master Seung Sahn, an effervescent man with a shaved head, gray robes, and a non-vanishing smile. He had established Korean Zen Centers around the world and written several books of playful Zen teachings.

Among the honored guests were Lama Chagdud Tulku Rinpoche who fled Tibet after the bloody Communist takeover; Cambodian Monk Maha Ghosananda, who survived the wholesale slaughter in his homeland and heads Cambodian resettlement around the world; I Ching Master Tan Ho of South Korea; author/activist Jack Kornfield; female activist Rabbi Lynn Gottlieb of New York; and Stephen Gaskin, the former Haight-Ashbury spiritual teacher and author.

These and other leaders gathered to endorse a World Peace statement to be sent to 250 political leaders around the globe. The document included several paragraphs and a single poem born from the unity of these minds:

The earth is spinning through space.
When clouds disappear,
There are 10,000 miles of blue sky.

Rabbi Gottlieb, her chestnut hair flowing to her waist, opened the conference by lending her clear mezzosoprano to an entrancing chant that invoked Hopi, Lapp, and Jewish traditions.

"Drop a stone in a pool," observed featured speaker Kornfield as the three-day conference opened. "How far do the ripples go?"

"The outside nuclear bomb is not so dangerous," host Seung Sahn proclaimed. "What is dangerous is the inside nuclear bomb: each person's like/dislike mind. If you let go of your conditions, situations, and opinions, then world peace is possible."

It seemed an appropriate gathering for this wooded nook in northeastern Rhode Island, the little state Roger Williams dedicated three centuries ago to religious broad-mindedness. It was fitting that East and West congregate here to inspire and create a matrix for peace.

The center had set a sumptuous vegetarian feast for a late lunch. I took the opportunity to collar Seung Sahn for a couple of quotes and to tell him about my bursting Buddha. He liked the story, or at least he laughed gleefully, pulled me by the sleeve, and helped fill a plate with great-looking goodies. As I was about to take my first bite of a luscious looking pastry, he thrust a wagging finger in my face and shouted, "No, no! Don't eat!"

I stood there with my mouth open as he continued laughing and walked away, leaving me, I think, with a teaching of some sort. I ate anyway.

"I'd guess he was telling you to let go of your broken statue," Stephen Gaskin speculated later that evening. Gaskin, who helped found The Farm, a thriving commune in rural Tennessee, attributes his own spiritual awakening, at least in part, to the appropriate use of the hallucinogenic powers of LSD during the 1960s. The Farm has gone through many changes since its origins in the early 1970s, but overall it has remained one of the more effective communities of its type. A former college English instructor, Gaskin writes and lectures on political and spiritual matters.

"You know, let go of the enticement right in front of you," he went on. "Forget the food. Don't turn away from the lesson of the busted Buddha. You can't afford to slip up. You have to be aware of your developing presence. After all, that presence is the passion that fuels you. Turn your busted Buddha into a trip. It's like getting high and then reading the teachings of the masters. That's when you can get it best. That's when you begin to really see it, to really hear it.

"There is a movement going on," Gaskin observed, "and you see major elements of it at this conference. The way you view the movement, whether in 1962 or 1982, depends on how far you can expand the deepest reaches of mind, how wide you can open your creative center. Some people see it as a civil rights movement or a religious movement. Those aspects are present, but the movement itself is much more powerful than the limitations implied in such descriptions. This is not just a Native American revival or some sort of commercial push with a 'New Age' label attached.

"In America today, there's an awareness escalation. You can't define it in limiting terms of social or political conditions, because the actual thing happening is that the world mind has spiraled into its own heart, and America happens to be at the focal point of potential. The center has to be somewhere and it happens to be here, in terms of economics and influence and overall power. You want a peace movement to work? Start it here. You want awareness to expand? Introduce the teachings here. This twentieth-century United States culture has the economic and intellectual resources–and, yes, even the compassion–to affect the whole world. Those at the center must accept their responsibility for the welfare of the whole and prepare to act when the appropriate time and situation appear."

This view of the modern world is reflected in the writings of many current philosophers and spiritualists. In *Manifesto! For A Global Civilization*, theologian Matthew Fox and physicist Brian Swimme state it clearly:

> And in this respect, Americans have a special responsibility in the creation of the global civilization itself. For in America there are diverse strands from the powerful human creation of our history: there are the scientific and artistic worlds that strongly extend and deepen the intuitions of the Greek axial civilization: there is the spectrum of religions, and the deep religious fervor and passion that we have been given by the great axial Hebrew civilization. There is the unique geographic situation where America touches the edge of both East and West. And even more, there is the great untouched resource of our continent's spiritual treasures; the form of life and being that is the Native American's. The North American continent is a most powerful locus of the creative power needed for re-creating the world.

"It's the old question of 'Don't just sit there, do something' versus 'Don't just do something, sit there,'" Jack Kornfield said during a break in festivities at the Zen Center. Kornfield teaches meditation, and, like several of the conference guests, his writing has become increasingly popular during the decade since the conference. "We are meeting here today to issue a statement on world peace. Statements on peace usually don't have much effect. That's a simple fact. What leads to people's actions is what comes out of their own hearts. One inspired person can do more than a million statements. Look at Gandhi, one committed person. Look at Mother Teresa.

"What each of us must perceive is the direct situation of the world. That situation is that there's suffering. If one person perceives this and acts, then that suffering is affected. It is not a question of ceremonies or religion or Christ or Buddha or politics or any of that nonsense. There are two sources of strength in the world: people who aren't afraid to kill, and people who aren't afraid to die. Each of us must do what we can, whether it's sitting in *zazen* or marching in peace parades in New York. Whatever, we simply must see the world more directly."

Later that evening, a panel discussion featured Seung Sahn, Chagdud Tulku, Maha Ghosananda, Zen teacher Sotetsu Yuzen Sensei of West Berlin, and Gaskin, who guided The Farm into developing its own international relief organization. The four-hour discussion was far-reaching, and featured the high humor of Seung Sahn and his long-time friend Ghosananda as spiritual comedians. This was no somber ceremony. At one point when the discussion threatened to sink into heaviness, Seung Sahn interrupted with the laughing admonition: "Have you seen *E.T.* , the movie? Just love. Just love. Get E.T. mind, OK?"

"I am in awe," Gaskin said to the panel. "I am sitting here with a Tibetan lama, a Cambodian monk, a Korean master, and a West

German teacher. I'm the only one on the panel who has not seen a major war in his homeland in this lifetime. We have come here not to agree with each other, but to improve the way we speak in order to learn to reach anyone and everyone who is the least bit reasonable. We must realize this problem of peace is not going to be solved in one generation. We now have to teach our children and our children's children. It's what I call the fifth vow of Bodhisattva: I vow to shovel manure against the tide forever."

"Good taste!" shouted Seung Sahn to the packed and laughing house. "Good taste."

Gaskin went on to speak about the many problems afflicting the state of Native America. He talked of acid rain, political injustices, squalid conditions on reservations, and the terror in which many natives of Central and South America were perishing. A decade later, only details have changed. Some progress has been made, but the complexity of problems remains as though our era could not proceed without them.

"Accepting responsibility for creating change in the world means being willing to move to the heart of not only your own creative center, but also to the heart of the world's creative center," Wolf explained during a meditation years after the Great Masters gathering.

"Does that mean I have an ancient heart different from the world's ancient heart?" I asked.

"Different, and again, the same. Just as heart and mind are different and the same. Your own personal hologram is a tiny part of the world hologram, yet the two are intrinsically aligned. You are obviously part of the world's total picture, and you have the capacity, through expanding your awareness, to shift your comprehension to the big picture as well. You can see the whole significance and insignificance and all the implication of the world's Twenty Count by mastering your own version's vision. To do it, though, you must become a master shapeshifter."

"OK. Explain that one."

"Through all our time together," Wolf said, "you have accepted me as a guide, but you have never seen a most obvious aspect of my presence. I too am an element of Twenty Count. I was with you the day your Buddha burst and your fantasy of religious bliss realigned. I am the guide of Five, separated self. I connect your individuated consciousness to all other circles in the Count. I'm with you at all times because I have to keep interpreting your interactions with all your other guides and numbers. The flexibility for this work comes to me only when I jump my awareness around in appropriate forms to disengage or at least disempower your daily fantasies. Otherwise, the fantasy level of your life would kill you in one day. You'd become so instantly deluded that you couldn't tell whether rush-hour traffic was approaching or departing. I must shapeshift to teach you, and you must shapeshift to send the teachings on beyond yourself. Guess what. You can use me as your role model."

"Delighted," I said. "Give me some guidelines."

"Just remember when and how you've seen me," Wolf answered. "For instance, I just told you that I was there on the day of the bursting Buddha. You accepted that, and that's good. But a few years ago, your first objection would have been that the Buddha fell long before you started your work with Twenty Count and its guiding spirits, and therefore you wouldn't have believed I was there on Buddha day. But you've now advanced to the point that the chronology of events doesn't dominate your consciousness. You're able to view the flow of your life in synchronous terms without losing the perspective that you're still a human in chronological existence on Mother Earth. That's your grounding as a shapeshifter. That's the point to which you can always return.

"Next, think of my forms. You've heard and seen me as Wolf, Fox, and Coyote. As a personal guide, as a flying form leading you into new realms, and as the tricky road of service to others. There's no limit as to how I can appear, but there will always be a clue that it's me. Next time, I might be a German shepherd, or maybe Lassie. Get it?"

"That's quite a trip," I observed to Wolf. "Letting go of individual identity to become a multiple being that is part and parcel of the whole universe right here and right now."

"It's an eternal matter of letting go," Wolf stressed. "Seung Sahn wanted you to remember the Buddha breaking, the crash, the release you experienced. The worst thing you could have done was to run out and buy another statue. He wanted you to stop eating so you would remember your contact with that original experience. Remembering requires passion. That means moving straight ahead when you let go of something, using the release as an implosion of spontaneous combustion. Life is a constant exchange, an endless struggle for balance. Humans get into the fantasy that life is more properly a process of accumulation and holding on. But that defuses combustion and destroys passion.

"Developing awareness of the ancient heart is the process of passion. That's the pathway. That's the wind blowing. That's growth. Focusing appropriately on your own development releases tons of passion through the love of moving energy. That, in turn, is what heals on all levels."

Wolf's teaching echoed the messages of the Great Masters World Peace Conference.

"The agendas of the hero's will and the perfect servant come into play here. These concerns must be balanced so that those at the controls neither abuse power through self-serving actions nor become victims of power's presence by assuming a powerless stance. A middle way, such as Twenty Count's process of moving into a flow and surrendering to sophisticated, proven inner guidance and change is the appropriate approach to this power.

"The world's Twenty Count is spiraling in on itself as the twentieth century concludes, and it's urgent for the creative center to respond appropriately. The old mythologies of nationalism—that's the world-level version of separated self—and fear-based manipulation must break down. A new imagination must arise so that everyone can see. These old mythologies are coming apart. The prevailing curse of the frozen intellect—the obsession of relying solely on information without converting it into knowledge and then wisdom—is coming to an end.

"A creative time is emerging as the wind of imagination realigns our mass fantasies into new perceptions," Wolf stressed. "It is a time of learning to envision and to speak. To listen and to grow. And do you know what all this envisioning and speaking and listening and growing is? It's the machinery for change. For you as well as the rest of the world."

"Yeah, that sounds right," I said.

"Think about it," Wolf urged. "As you work your way through Twenty Count, there are four major steps: the realizations of separated self, expanded awareness, kaleidoscopia, and the return to the ancient heart. Those are your plateaus on the path of the mystery game, the rungs on your ladder to freedom."

Behind closed eyes, I saw the four circles of Five, Ten, Fifteen, and Twenty emerging from a ghostly spiral at the creative center.

Within each circle stood the vertical black diamond showing the upright human reaching out to holographic access of kaleidoscopia and the ancient heart. Wolf, or perhaps Fox, was about to conduct a tour around these circles. In my mind a neon sign flashed: *Welcome to Coyote Road.*

An overwhelming loneliness set in. I was isolated, removed, alienated from the rest of existence by time and space in the midst of a barely perceivable spiral. Emotions flowed like water and thoughts sprang into existence fueled by the emotional sensations. Self-focus translated directly into self-importance and self-pity, highlighting Seung Sahn's nuclear bomb within. I had to fight for my own rights, brother. Nobody else would do it for me.

Fortunately, growth abandoned this childish position and propelled me into expanded awareness. There, faith eased the internal threat of alienation and nurtured my various dimensions. I accepted the existence of right and wrong, and committed to doing things the right way. My plaster Buddha had been my ideal of the right way, my religious position. It had felt good to tie my identity to the movement and balance of north. It felt like home. Then it all exploded.

The bursting Buddha boosted me into kaleidoscopia. Releasing my conviction of an absolutely right way to live opened up a new world of potential. Had I held on to my image of righteousness–my eightfold path or whatever–my worldview would never have opened to kaleidoscopia.

"Forget your history," proclaimed Holygram the hippie amid floating psychedelic images. "Kick back, man, and work on your sacred dream. You can realize it here in kaleidoscopia, but you have to be pretty cool to project it out into the material world. Trust me on this. You're not quite ready yet."

The next leap was through the gateway to freedom directly into the vision of the ancient heart. Gone were separated self, expanded awareness, and kaleidoscopia. The ancient heart alone existed. I could hold the focus only with the greatest possible diligence. I realized that even carrots and bananas could be major distractions and destroy my fragile perception and balance. My head pounded and I broke into sweat as I struggled to maintain the opening.

"Come on back," Wolf said. "Relax, relax. It's not time for you to go into the ancient heart yet. Now is time for you to go out into the world. But you must realize that the point of your active outer life on Earth is not to run around babbling about who or what you think you are, but rather to become a clear, balanced functioning being with your feet on the ground, your mind focused by instinctual intellect, and your actions precise and effective. If you get lost by obsessing on diddly personal or religious or political concerns, your imagination simply will not be fully focused on creative straight-ahead possibilities. You'll be lost in personalities and fantasies again.

"For instance, though you like to say you have walked a Native American spiritual path, that terminology describes nothing more than your religious observances into and through expanded awareness. There's a whole generation of Americans obsessed with drums and rattles and sweat lodges. And for that matter, a generation of Native Americans who are squabbling about whether or not it is appropriate to teach and write and sell such things. This is all about religion, not about the movement of spiritual power into and out of the creative center.

"As you live the rest of your life, you will continue to grow only if you move beyond the need for identifying with any idea or group or country or whatever to keep your mind open and functioning. This is not a simple leap. It's the trip through kaleidoscopia. You've got to use your mind in a very special way to accomplish this. But I have faith that you can do it. Deep down you've got possibilities."

"Are you sure?" I asked. "Sometimes I'm not."

"Cut the crap," Wolf answered. "That was just your own little one-sentence trip back into separated self. I'd think you would get tired of that place."

"You're an amazing guide," I said. "You're an inspiration and you also know me so very well. You won't let me get away with diddly. You speak to me on a level even deeper than Old Ghost, or so it seems. Is that true?"

"Of course," Wolf answered. "I know you completely because I have always been part of you. My eyes were the eyes with which you first looked out onto this world. You have heard my howl since the instant of your primordial flash. We are not separate. You see, I am your soul."

KALEIDOSCOPIA

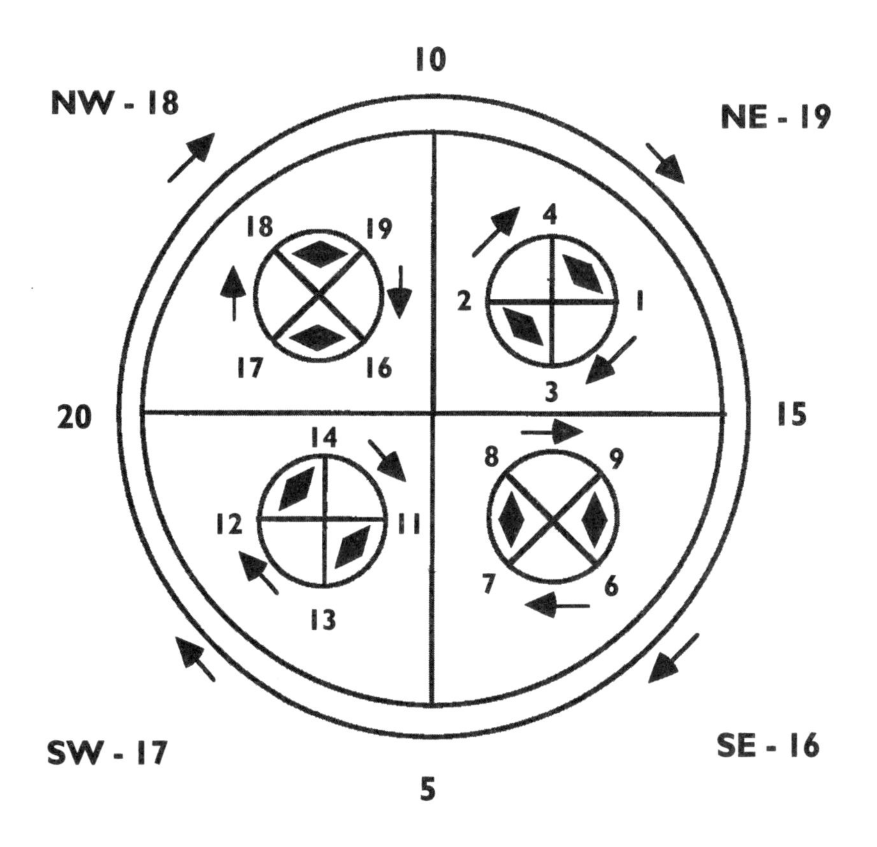

NE Tumbler 1 to 4: Chaos

SE Tumbler 6 to 9: Order

SW Tumbler 11 to 14: Hero's Will

NW Tumbler 16 to 19: Perfect Servant

Grand Tumbler 5, 10, 15, 20: Passionate Detachment

Tumblers

I hummed into a deeply relaxed state. The yogi stepped behind me and placed his fingertips on my closed eyelids. He said he would give me a magical gift. Gently at first, and then more firmly, he pressed. The steady pressure grew hot, and the black inside my eyes turned red. Something popped painlessly. Just "pop." I saw stars, lots and lots of stars.

Today, years later, I close my eyes, gently press them as I breathe in a particular manner, and a universe appears: a great black vacuum, its foreground populated by stars, planets, and geometric forms. The vacuum divides into left and right ovals that merge into a ghostly, whirling wheel called the grand tumbler. Long ago, far away, the grand tumbler spins into colors and splays into fractals.

As I maintain the gentle pressure on my eyes, the grand tumbler whirls back into darkness. A vividly defined eye evolves, complete with shimmering pupil. Within the pupil, movement. The movement of smaller tumblers revolving within the grand tumbler.

Wolf explained it this way. "You're seeing kaleidoscopia, the holographic field that displays all elements of past, present, and future. This is Fifteen, the hallucinogenic world of Holygram. Inside the grand tumbler, four smaller tumblers rotate clockwise. Each tumbler contains four segments. Two contain upright diamonds, the shape of expanded awareness reaching out to kaleidoscopia and the ancient heart. The other two segments house the invisible forces within all beings.

"The four tumblers are the hidden agendas that beset human beings. Starting in northeast and moving clockwise to northwest, these agendas are chaos, order, hero's will, and perfect servant. To break though your mental barrier, you must figure which hidden agenda is motivating you at a given moment. Then, moment after moment. Only when you disable all four agendas do you develop the freedom to look straight into and through kaleidoscopia. That's when you see the ancient heart."

The little miracle of the tumblers vision lives within me, behind the eyes, constant evidence of my limited comprehension of existence. Nothing before or since the yogi's fingertips on my eyes has revealed such a complex vision. Adventure steered me to this shamanic gift, then the gift forbade me to close my vision ever again.

The implications of this vision deny the core of many belief systems we use as starting points for thinking about existence. If the universe exists within, why exhaust ourselves seeking answers in the outside world? Faced by the ruthless demand to choose between my tumblers and a personality-centered belief system that showered grief and pain relentlessly, I selected direct perception, the path of no discernible resistance. Staying in touch with the base of my altered thought core emerged as an all-consuming challenge.

That base, of course, is Twenty Count. Maintaining focus on the spiral nature of life provides a perfect antidote for hidden agendas. Living the spiral means practicing passionate detachment, at least partially through an ever-present sense of humor. Twenty Count requires, at every moment and in every situation, the ability to step back from whatever involves or obsesses us and laugh a deep belly laugh at ourselves. This act alone disables many hidden agendas, at least for a moment. The power of such laughter makes itself available through the big picture–the full view of kaleidoscopia–by revealing every person as a multifaceted being who can become a comedic fool by obsessing on any of the hidden agendas.

To achieve this big picture through the grand tumbler, it is necessary to detach identity from personality. Peering into the small tumblers forces you to retrace your core beliefs, if possible, all the way to their source. I tried this retracing by reflecting on childhood memories in meditation. I relaxed, breathed, pressed my eyes and watched the cosmic show, then located my moment of birth.

My first perceivable sensation registered as pure pulsation. After a while, I identified an image, perhaps my mother's face, with clearly defined boundaries. In time, I realized she was separate from me. So closely bonded were we, though, that I saw what she saw. Still, my vision was my own. I projected an outer world, which, as I learned to sit and stand, came into balance as reality. I fell in love with this reality.

I became educated by identifying creatures and things around me. As a side effect, I also became alienated, losing my babyhood sensation of life as a whirling blob of liquid energies. I created an image of myself by blending a hodgepodge of projections into a mushy, slushy concoction called personality. This I offered as my shield against a world I saw ruled by harsh sensations and alienation. As a result, my every perception of life and my every action toward the world and its inhabitants became reactions programmed by my shield as protection for my self image. My shield grew so powerful that it developed into a four-walled personality fortress, isolating and imprisoning my true sense of self in its dungeon of denial. My love of reality had come to this.

No person or process alerted me to reintegrate my identity with the original pulsating energy, and that vibrating essence vanished from my memory. I identified with my physical body and its reactions. I was convinced that my programmed and shielded reactions–my personality–defined my start, existence, and end.

At some time, I sensed something was wrong and started searching. I discovered that for humankind as a whole, releasing our trapped identity from its personality prison was the goal of all spiritual disciplines. It differed only in form and method.

In Zen, a meditator sits silently for hours at a time, watching for the creative center to flash its next light sensation. If a golden Buddha appears, the master says just keep breathing and it will go away. Stop identifying and just see what's there. Go directly into the void.

In yoga, a powerful energy called kundalini is released from the base of the spine through a system of energy centers along spinal pathways to enlighten both inner and outer being. Acupuncture's body meridians correspond to lines within an auric field that healers read. Fakirs and ascetics through the ages have practiced esoteric disciplines to develop these energy perceptions and translate them into realities. Their teachings have reached a wisdom-starved audience in the Western world in recent years.

The purpose of working with the energy body is to destroy the fantasy that personality is the same as identity. The mind starts its Earthwalk by attributing unconditional reality to the body and the personality. But if mentality can expand enough to perceive the energy body as equally real as the physical body, then the mind's fixation on body and personality loosens its stranglehold. New possibilities arise to pave a path for expanded awareness. The more that we can balance the obsessive focus on the physical body with the imaginative potential of the energy body, the closer we move to direct perception.

A different model of an energy body has been provided for current seekers by Carlos Castaneda's writings during the past three decades. Through the teachings of his mentor Don Juan, Castaneda has introduced millions of readers to an egg-shaped fluorescent oval formed by a human mold and functioning through its own inner technology. Castaneda's model is presented through the development of an innovative vocabulary in nine books. An entire generation of readers and writers now borrow from Castaneda when referring to energy concepts. He has established new and expansive meanings for terms such as "apprentice," "impeccability," "stalking," "dreaming," "assemblage point," "sorcerer," "eagle's emanations," "*nagual,*" "*tonal,*" "personal history," "death as an advisor," and others. The effect of this vocabulary has been the acceptance by many seekers of a clear-cut image of the glowing egg-shaped oval as a new energy-body alternative.

Castaneda's presentation thus shares purpose with many another discipline: to release identity from its walled-up personality prison. The ever-present mental barrier is, from this angle, nothing more than the personality, and our hidden agendas are enduring fantasies whose purpose is to project and protect personality. Disabling these agendas leads to rediscovering our spiritual identity.

My own journey of introspection through the tumblers was facilitated by five guiding spirits. The first was the spirit of Ten, expanded awareness, representing the grand tumbler.

Grand Tumbler (Five, Ten, Fifteen, Twenty)
Art of Passionate Detachment
Guide: Scribe, Spirit of Ten

This spirit was Scribe, an artist whose works went up in smoke long ago. He was an ancient Maya, his face mirroring the slanted forehead and prominent nose that sculptures and other artworks associate with his people. His rising topknot of gray hair blended into smoky clouds where his wisdom has been stored. He embodied the art of passionate detachment.

"You can see that the numbers Five, Ten, Fifteen, and Twenty are included in the grand tumbler," Scribe explained before we even were introduced. This being of heart and intellect had no patience for a meeting of personalities. "These are the numbers of the progression of self from the first realization of separation, to the second realization of expansion, to the third realization of inclusion within the holographic universe, to the fourth realization of returning to the source. The human being who can maintain awareness of this perspective throughout every day and night of existence controls the process that opens the pathway to the ancient heart.

"The fantasies that infect the mind of human beings are based on thinking that you yourself are the primary entity in the universe. You can see, however, this is not the teaching of Twenty Count. In the Count, primordial awareness is One, and separated self does not appear until Five. Fire, earth, water, and air all precede humanity in existence. If not, humanity could not survive. This simple truth is the basis for dismantling all hidden agendas. So, it's time for you to peek into the four small tumblers and examine their hidden agendas once more. This time, you will meet a spirit guide to illumine the avenue of release from each agenda."

NE Tumbler (One to Four)
Hidden Agenda: Chaos
Guide: Shivering Turtle, Spirit of Nineteen

In northeast, Scribe pointed out, the hidden agenda of chaos dominates because of our blind faith in perception of ourselves as personality. Though persisting in this agenda allows us to plow a path to victory sometimes, we also are plowing afresh the fields of nurturance for ties that blind. Everything we do impacts other persons and situations, and often we're unaware of these implications. Chaos, simply, is the condition of not knowing ourselves and unconsciously spreading the results of that not-knowing like seeds of poison ivy along every path we walk.

The guide of Nineteen was Shivering Turtle, the guardian of the gateway to freedom. He surfaced as a pure, resonant voice accompanied by ripples of energy up and down my spine. In contrast to the sparkling essence of this electrical flow, however, Shivering Turtle repeatedly counseled constant patience, vigilance, and endurance.

"Like the turtle," Shivering Turtle's voice advised, "advance slowly and wear the spiral mandala as a shield on your back. You will discover that the easiest way through all this complexity is the longest way, the path that winds around and around and just keeps going until it returns to its source. Embodying and passing on the wisdom of Twenty Count has become your true quest. You started out crying for clarity. That was your fantasy, your glue to your personality. But everything was clear all along. All you needed was a smell for a real purpose. Now you are meeting us, the guides. We know it's difficult for you to live without agendas, so we help you muddle through. You are much more familiar with all four of us than you realize. We're always ready to work. Living on your other side leaves us a great deal of free time, since we are not bounded by your many distractions and since you cannot spend more than a tiny portion of your waking hours beyond your mental barrier with us. It would be easier on your learning process if you could. Listening opens you to transcend both coyote road and fantasy road. That's the goal, both within and without."

SE Tumbler (Six to Nine)
Hidden Agenda: Order
Guide: Lama, Spirit of Sixteen

The hidden agenda of order includes numbers Six through Nine and emphasizes awareness of history, dreams, relationships, and decisions. This is a world where behavior is prescribed as a list of shoulds and musts, the mental garden where philosophies and belief systems sprout and bloom. Duty and honor are valued attributes, and spontaneity bears the burden of a governor on its shoulders. The world of order provides strong reference points for building and guiding institutions and organizations, but it severely limits growth beyond the boundaries of its own system.

My guide for Sixteen was Lama. A Tibetan elder, his image overlapped my memory of Chagdud Tulku Rinpoche from the Great Masters World Peace Conference and my imagination's projection of T. Lobsang Rampa, the English author of *The Third Eye*, a metaphysical book that blended Tibetan images with the writer's inner perception of his embodying the spirit of a Tibetan lama. My own Lama was a powerfully built man with chin whiskers and a most serious mien. He spoke to me late at night in alpha state before dreams dropped in.

"You have asked how to get rid of unwanted things in your life," the wise Lama mused at our first session. "The true and valid question, however, is: Why not give everything away instantly? Surrender personal identity as you promised up Algonquin Peak, and historic entanglements will vanish with your identity. Realize that you, like Wolf, are really part of a whole. You are light, and light is simply vibration. You are part of all people, working together and living together. You are a family, a community, a nation, a world, a universe. Realizing this, you will manifest an incredible opportunity to empower other beings by eliminating personal shoulds and musts. And that, of course, empowers you as well. That's the other side of personality. At this point, you will be able to grow any crop and operate any computer. In other words, you'll function in reality rather than dysfunction in fantasy."

SW Tumbler (Eleven to Fourteen)
Hidden Agenda: Hero's Will
Guide: Dream Healer, Spirit of Seventeen

The agenda of the hero's will is located within the progression of Eleven, Twelve, Thirteen, and Fourteen, where expanded awareness organizes and transforms itself into instinctual intellect. Awareness and confidence can propel the individual into healing and improving the human situation on Earth, but it can also implant a savior complex as well.

Dream Healer was my Seventeen guide. She seemed to shapeshift like Wolf, but her image was always a wavelength away from my fastest eye-shift. To receive her communication, I dreamed. When I woke, I recorded her words.

"If you can hear me in your dream world," she whispered, "then you are making a successful transition into synchronized living. This means forgetting the ideal of separated self as hero and instead submitting to the will of spirit. You will be more creative, more inspirational, physically stronger, and, oh yes, you will develop a wonderful personality."

The dream suddenly verged on nightmare. Dream Healer was holding up personality as a goal in the midst of my newfound mission to destroy personality. What was going on?

"Now, now, don't be shocked," Dream Healer continued. "No word or concept is either good or evil in itself. You are so literal. It's all a matter of the spiral power of the mentality that uses that word. You've examined concepts like 'bliss' and 'fantasy' in terms of Twenty Count, but you'd do well right now to go back and reexamine them. Those you've seen as negative, see as positive, and vice versa. Light as well as shadow inspires fantasies. As you delve further into deep mind, you will indeed discover and develop a delightful range of personality."

In the lighter-hued and more-elusive dream that followed, I remembered that I keep forgetting the powerful face of Red Thunder Cloud, the man who was always laughing. What a personality.

NW Tumbler (Sixteen to Nineteen)

Hidden Agenda: Perfect Servant

Guide: Karma Master, Spirit of Eighteen

The perfect servant agenda encompasses Sixteen, Seventeen, Eighteen, and Nineteen. Lama, Dream Healer, Karma Master, and Shivering Turtle live in these numbers and set wonderful examples for humans. They are role models in their selfless service of awakening us from our fantasized identities.

"It's all a matter of practicing the art of passionate detachment," said Karma Master, the guide for Eighteen. "The power of laughter is a great tool for distinguishing the possessed from the blessed. The grand tumbler employs the big picture, allowing a full view of the four agendas and their potentials. Perceptual fantasies can be extremely effective power sources. Kundalini is fantasy and so is the energy body, but they can produce very real effects in the physical world. After all, are not we guides ourselves fantasies? Of course, we prefer to be called myths, but we're not attached to it. Our own myth is that we're passionately free of all labels.

"You must understand," Karma Master continued, "we guides can live in our mythological state because we're evolved as masters. For the same reason, we can work selflessly for you, guiding you literally forever. But if you do the same thing for someone else, you run the risk of turning yourself into a slave. We do not have individual bodies and minds and personalities to maintain. You do. We do not want you to emulate us, but to learn from us.

"That's why there are elders on your planet. Their purpose is to keep the children from becoming slaves by teaching them to think-sing. Elders can think-sing children into seeing their individual tumblers within the grand tumbler in order that they do not become so attached to their agendas that their passion dies. Just as there are different segments in each tumbler, there also are multiple segments in each human, including an elder segment given to the reality of experience and a child segment given to the fantasy of imagination. The only way elders and children can align with one another is through silent think-singing, a form of communication that conveys an inherent mythology. That's why elders must be honored. They hold the keys to unlocking the mysteries within the map of hidden secrets.

"Elders are not myths, you see. They just remember them."

SILENT SELVES

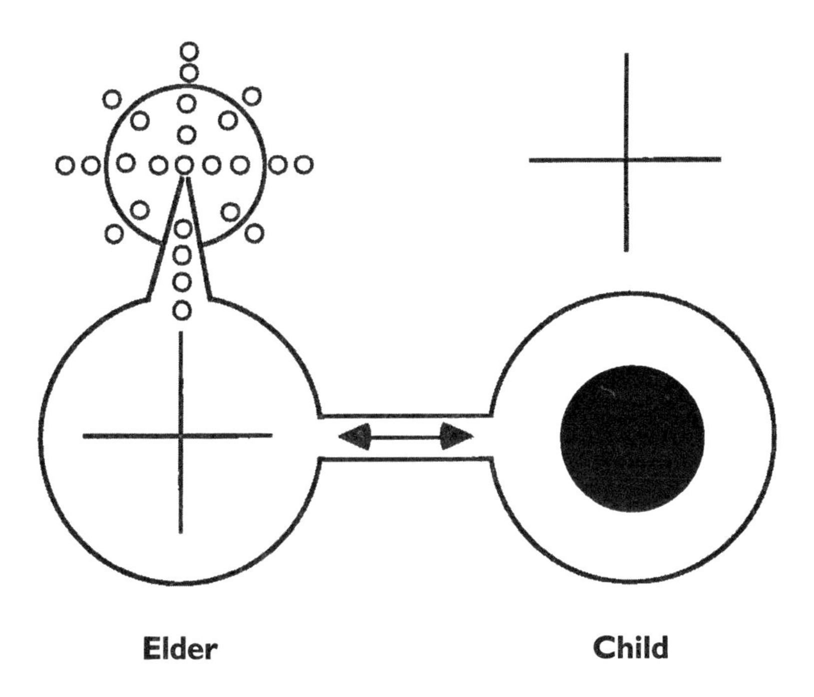

Child, focused inward.

Elder, wisdom from above.

Two Faces of Age

At the heart of the flash, color and song dominated my every sense. Rather than a separated sentient being, I existed as a collaboration of eyes and ears. The silent selves appeared as twins, promising to explain appropriate life roles to adopt and embody during this Earthwalk. Their voices blended in chanting unison. Though seemingly identical, they projected differentiated perspectives. One was Child; the other, Elder.

"We are the two faces of age, the solar and lunar aspects of original flash. We are both teacher and learner. We dwell within Twenty Count. Separately, we are alpha and omega, first and last. Listen to us and learn, then teach us and watch us expand and grow."

Child was the spirit of Three, the first projection of humankind, the blank slate born from perception and reflection as a product of the resolution of contradictions. Elder was the spirit of Four, a balancer of the cross within the circle, an experienced being with in-depth understanding of how the order of chronology blends most appropriately with the spontaneity of chaos to create a dynamic existence.

"We speak to you, both individually and separately," they chanted together silently. "We are both part of you. The more you can incorporate us into one another, the better for you. If you can apply the wisdom of age to your youth, you will make powerful choices. If you can retain the spontaneity of childhood into old age, you will enjoy the enlightenment of experience. If you can maintain awareness of this guiding principle, you will discover that your present trip through Twenty Count is taking place on the other side of the creative center. You have passed through the ancient heart and are experiencing the reflection of your previous tour of Twenty."

Frozen silent, I existed only in the midst of primordial flash, hearing and remembering. The silent selves appeared as circles. In the east circle, Child's undeveloped intellect focused inward toward the mystical darkness of perceived pain and confusion through which it had recently sprung, while the four-directional cross hovered overhead, its processes and messages momentarily out of reach.

In contrast, the cross of Elder was centered in west where the introspection of age rules and the comfort of wisdom releases the anguish of youth. Above Elder hovered Twenty Count, and the two were joined by a ray from heart of Count to mind of Elder. Child and Elder were connected by their own heart-to-heart passageway through which the wisdom of ages flowed freely in instantaneous dialogue. Though their medium was silent song, it evolved into conversation for my perception. Enraptured, I clung to every word.

ELDER: "Ho, Child. Welcome to Earth. I know you are disoriented, and I am here to comfort you. You feel strange here, I know. Perhaps frightened, ungrounded. So listen quietly to my words and you will begin to know yourself. You may not understand all I say at first, but in time you will value what I tell you. Your biggest challenge will be to remember me and my teachings. You see, I am a completed being. My years have centered me. I no longer believe that I am just a name or a body or a personality. My intellect is concerned with, and focused on, higher matters. I have discovered many secrets imbedded in my mind, and it is my responsibility to pass them on. I must send a new image of the medicine wheel and its teachings forward to the coming generations. If I do not send this message about this light and shadow show, then our children will lack a grasp on the appropriate flow of living. The techniques for discovering and utilizing compassion, intelligence, and power will have vanished. But I can deliver this wisdom only if an open mind is prepared to receive it. The message needs a perceiver."

CHILD: "Ho, Elder. This is my first thought: I am that open mind. I am unformed. Your challenge is to teach me. My responsibility is to maintain the vision of the ancient heart and receive your teachings. I am ready. Beyond that, I think nothing. But I feel, I feel so many things. All these vibrations blowing about so wildly. I think I am afraid. I feel I should know things, but I don't. Why do I know nothing?"

ELDER: "Your question is good. Your first teaching is about remembering information, nurturing it into knowledge, and reaping its rewards as wisdom. Information is the light and shadow contents within the field of vision in front of your eyes, and you may feast on it at will. Knowledge is utilizing wind and order to align the light and shadow vibrations. Wisdom is the passageway of the heart to the mind, the tunnel that connects to the mystery. It's like a flute. Our purpose in life is to make music on the flute. Our lips and breath and fingers maneuver the tones to produce the song of life. But to make your music, to learn to sing with me, you must first remember. So breathe and close your eyes and call back your primordial flash."

CHILD: "Yes, I see the flash. This is fun. I am flying through the air on the back of Fox. I cling to the body and feel the heart beat within me, and I know what I must learn. Please, Grandfather, teach me the games of Mother Earth, the thrill of bursting out of my dream world through my own muscles and coordination. I am full of skills, and I must enter competition. But I will compete in order to master myself, not simply to win a game and remain a victim of my mind."

ELDER: "This I can teach you, but all in the appropriate time. First, you must awaken to the world around you, not just to my voice within you. You are still a newborn, and you must grow in body and learn the possibilities and limits of your mind. At this moment you are dreaming, so come now, come to the light."

CHILD: "Ah, it's blinding. So now I wake up. Grandfather, please help me. I am fragmented. I stare around this world and it seems a maze of unidentifiable liquid energies merging into one another's colors and sounds. All I remember is darkness. I have no control over this world or this body I now inhabit. Please, Elder, help me. Now, I need you. I don't know where or when I am."

ELDER: "I will teach you, Child, one lesson at a time. There is no need to rush. As you are learning, there is pain when the four directions of teaching are not grounded within the circular flow of your existence. While you are a baby, your intellectual faculties remain undeveloped and you cannot master control of your body and mind processes. As you start to see what's happening in your body and on your planet, a significant question concerns who you are, your identity. You look inside yourself to see what you can see. If you remember nothing but darkness from your trip out of the source, your mind will begin fantasizing immediately, and unless someone like me is present as a guide, you will establish a belief system based on childish fears and ungrounded thinking. Your life will become a vicious circle of experiences repeating themselves over and over in order to validate your belief system."

CHILD: "That sounds awful. You mean that after this birth trauma I just endured, the rest of my life may just be a rerun?"

ELDER: "Not necessarily. Remember, I gave you two alternatives. One, maybe you can remember something of your source, something besides darkness. And two, you have Elder–me–always here to teach you. All you need do is pay attention, and I will nourish you to the point of mental functioning in your body on your planet."

CHILD: "Great. So tell me one thing. Who am I?"

ELDER: "You are observer. It is just that simple."

CHILD: "Nothing's that simple. That doesn't even sound like an answer. What are you talking about?"

ELDER: "You are observer. You are the being behind your eyes and looking out through your eyes now as Child. The same being who was behind those eyes at the moment of birth, the same being who will look out when you are twenty years old, fifty years old, and just before your death, whenever your deep memory allows that to occur. And also, obviously, the same essence that vacates your body at death."

CHILD: "I'm beginning to see what you mean. But I'm just

watching? Nothing more? Sounds so passive."

ELDER: "Precisely. You are observer, and the only action you are capable of performing is perception. To whatever degree you can learn to control and manipulate your perceptive capacities, you will open yourself to the incredible and unlimited possibilities of this life you have recently entered. This is your alternative to the repetitive cycles of a belief system based on pain and limitations. You become passive if all you do is submit to your cycles of frustrating repetition. True action requires incredible concentration of your whole being, not just the flailing about of arms and legs while the mind glazes itself over with denial and desire."

CHILD: "All right. So I accept you as my teacher, and I accept myself as observer. What next?"

ELDER: "You grow, in mind as well as in body. You don't have to learn to grow, you just grow. Your intellect will stretch out to the world, reaching east for kaleidoscopia and west for the vision of the ancient heart. This pulls you up and out of your self, and displays your center as a black diamond, the image of the human. You will see that your alignment is vertical, and you will realize that the answers you seek about life—these answers I am giving you at this moment—are the secrets hidden in the folds of mind within that black diamond. The more you focus on the holographic nature of your existence as evidenced by both kaleidoscopia and the ancient heart, the more inner space you create for the cross with its four directions to imprint on your awareness. This is the process of letting go of the idea of an alienated self—personality—and embodying the fullness of growth, pure and simple."

CHILD: "So as I grow, there are specific things you can teach me, yes?"

ELDER: "Of course, many things, depending on the time and

space you have chosen for your body, and how my wisdom can best benefit you in that time and space. I will show the nature of duality's illusion and unification: for every poisonous herb that grows in the forest, an antidote also grows within life's reach. You must learn to observe. See the rhythms and polarities of life on its many levels, how every aspect of both you and the world reflects the nature of every other aspect. You must balance in order to avoid obsessions, which will blind you to your true nature. To balance means to bring in the teachings of north, the cross that hangs above your circle of life. You must draw down the teachings of the directions, your own personal Twenty Count, in order to become an adult. Within your Count's realms you will discover worlds of nature, art, music, business, love, war, possible accomplishments, and probable defeats. All of these things you must balance."

CHILD: "If I do this, if I draw down my medicine wheel teachings, will I eventually become responsible for myself? Will I be able to govern my own nature and perceptions?"

ELDER: "I sincerely hope so, though there are literally billions of humans upon your planet who still live as needy children. They have not grown into their realms of responsibility where they can apply the teachings. The culture you live in offers few if any rites of passage from childhood to adulthood. So you must remember yourself and create your own rites."

CHILD: "How do I remember myself?"

ELDER: "Recall how your perception of this conversation started. It began when you were asleep, before you opened your eyes and felt the sting of first light. It is through the dream world that you contact your deep memory. Within deep memory you will discover the rites of passage from the Twenty Count of the cosmos. Perhaps you will find vision quests or dances or music or martial arts. Turn your back on no possibility. Just know that whatever your rite of passage may be, you must understand it in terms of deep mind, with the full observation of the fantasy road and coyote road of your Twenty Count. You must meet and befriend all the spiritual essences within you. Only then will you possess the resources to change your own actions and thinking and personality. You will begin to experience body sensations of developing guidance, and feel the progressive changes in your muscles and bones. Your appropriate rites of passage will appear before you. But you must be alert. Otherwise, it is very easy to ignore them in favor of living the fantasy life that your culture will attempt to sell you from your first instant of worldly perception."

CHILD: "So what happens in my rite of passage? Do I change significantly?"

ELDER: "Yes, you will change. Your unspoiled, innermost nature of innocence and learning accompanies you through childhood. But you must release this child nature or at least balance it effectively to become an adult. You must begin to apply the lessons you have learned in order to pass on the teachings. Therefore, you will find yourself becoming more serious, tending more toward right/wrong fantasies, and examining philosophical issues more deeply. In order to avoid becoming embroiled in such weighty matters, you must discover the playfulness of life through your rites of passage. Then, you must hold onto that playful image, because as an adult you will function in two roles and both require a sense of humor. You will become a teacher and a teller. You will be a teacher to children, whether or not you physically are a parent, and you will be a teller of your lineage; that is, someone who passes on the essence of your Twenty Count, whether you call it that or not."

CHILD: "Can you explain these roles to me?"

ELDER: "Of course. As you grow into adulthood as man or woman, you will acquire two distinct types of knowledge. One addresses getting along in the physical, material universe of everyday life. This is your outer behavior, the actual work you choose, the lifestyle you adopt, your sexuality, computer training maybe, how you dress yourself. The second type is the hidden knowledge that you must recover from within your own deep memory. This second type of knowledge will provide guidance that will determine the inner approach you bring to your outer behavior. If you are successful in learning enough about yourself, you will focus on the expanding richness of the inner self rather than on the superficial outer self. Then, when you are older, you will pass on what you've learned. As teacher, you will set an outer world example for someone. You cannot avoid this role. Even if you never become a physical parent or live hidden away in some dank attic, that in itself is a teaching, and someone always is watching or somehow affected. Your absence from someone's life can affect that person. Teacher you will be.

"To be a teller, however, you must apply yourself diligently to comprehend, embody, and then re-form the inner teachings you have received. Your Twenty Count must live long, stew within your own matrix, and prosper before you can pass it on. When you are ready to tell what you have learned, the opportunity will appear. Of that you can be sure. You will never know exactly how to prepare yourself for this particular experience, though. It will just show up.

"Teacher and teller are two very different undertakings. Teaching demands ruthless patience. The teacher must never identify with the student's blockages and impediments even though they may be obvious and familiar. Gentleness is necessary in the face of the student's difficult inner struggles, while harshness may be the only answer to apathy. Through it all, teacher must remain unyieldingly calm and pleasant inwardly. Outwardly, it doesn't matter what the student sees if the learning ultimately permits the student to penetrate the applied manipulation of the instructional method.

"A teller, by contrast, lives in the vast, wholistic world of Twenty Count, traveling a spiral path and issuing bits of information to whomever passes by. When appropriate, teller passes on teachings about accessing knowledge and nurturing wisdom. But teller never becomes concerned with whether the recipient of such teachings reaches a finite goal. The spiral journey goes on and on, and sooner or later, every seeker returns to the creative center. The timing of that return is not teller's concern. Indeed, teller has a far different focus to maintain."

CHILD: "You are a teller then, aren't you, Grandfather? I see that. But what is this other focus you speak of?"

ELDER: "Yes, Child, I am a teller. And my focus is very clear. My focus is death."

CHILD: "Death? But why? Now that you've learned so much about life, why don't you concentrate on passing on even more knowledge to more children? Why concentrate on death?"

ELDER: "Because I can see so well what's coming. You are still a child, and even though you see the teachings ahead, mortality is not your concern. The pain you experienced arriving here has convinced your mind of your absolute nature, and it will take years for you see through your illusionary patterns. But I look at you in terms of Twenty Count and I see that you are the sea, the ocean. All life on Earth sprang from the ocean. I see the full cycle of birth, life, death. You are Child, still so close to your center that you sense the on-going flow of your existence despite the fantasy your mind is attaching to your painful arrival here. But your fantasy is influencing your interpretation of the birth/death experience, which you remember as the singular experience of crossing through the creative center. Death, with its anticipated agony that corresponds to the birth pain, registers in your mind as a happening to be denied or at least avoided. In short, your obsession with identity blocks your view of the true nature of death. But I resolved my identity long ago, and I do not fear death. Instead, death is my comrade, my comfort for life."

CHILD: "I do not understand. In a way, perhaps, but not really."

ELDER: "Death is nothing more than passing through the vision of the ancient heart and emerging on the other side of the wheel. We do it with every pulsating throb of our hearts. Death is the process of the shadow overtaking the light as we reconnect to our totality, of becoming One. One day you will meet a being called Wolf who will become a great friend to you. Later, death will be merely your process of becoming like Wolf."

CHILD: "I have a good feeling about Wolf. I know we will be close and I can't wait. So you are saying that death is not a bad thing? Not to be feared? Even by a little child like me?"

ELDER: "The quality of your death will determine the value of the ancient heart at the creative center of your Twenty Count. You do not know it yet, but you are playing a wonderful, powerful game. You are circling and spiraling, and you are constantly faced with choices. The more clever and aware your choices are, the better your game will come out. If you play just right, you will become a master. No, my young friend, death is not to be feared, especially by a child like you. You are preparing for your death with every breath you take. If you breathe with a clear mind and an enlightened spirit, you die the same way. And that is the best way to die. The shock of clear perception, precisely what we're preparing for during this life, will arrive at the moment of death. If we're interested in retaining the awareness necessary to go through the death process with consciousness, we must learn to open the creative center where we process clear perception. Otherwise the shock will be too much, and one more time we'll return to the primordial flashing to await the next shock of exit."

CHILD: "Wow, you mean life leads to death which leads to life."

ELDER: "You see it everywhere, in the leaves and in the stones. In time, however, you will discover that riding death with awareness into the spiraling ancient heart is the ultimate adventure of life, requiring clarity and courage beyond imagination. Otherwise, you get pulled back into the fantasy road one more time. Your spirit is a supreme tool for manipulating life and death. It is not a matter of escaping life, but rather using death to awaken your awareness. We have the option to embrace humanity and Earth and live to the fullest, or to hide within our self-imposed walls of limitations and blindness. Life is the time to use your breath. You don't take even one breath with you to the other side. Therefore, memory has no means on which to cross over. Not unless you befriend a guide who will ride you across in full fearlessness with your eyes wide open. For you, Fox is such a guide, and befriending Fox would be an act of great power. At the right time, you'll do it."

CHILD: "But what happens on the other side? What is death?"

ELDER: "You and I switch roles. You see, you just recently abandoned the other side in order to return here, so your memory of that side is much fresher than mine. Child emerges from death to inspire Elder to look back at life and remember. The process of shadow to light to shadow to light, our rhythmic dancing on the planet. When you become truly proficient in the dance, a strange thing happens. Shadow transforms and suddenly illuminates you, and all fantasies become instantly evident. You then vacuum the shadow in order to obliterate fantasies and, in their stead, manifest your cornerstones of identity. That is the purpose of your Earthdance."

CORNERSTONES OF IDENTITY

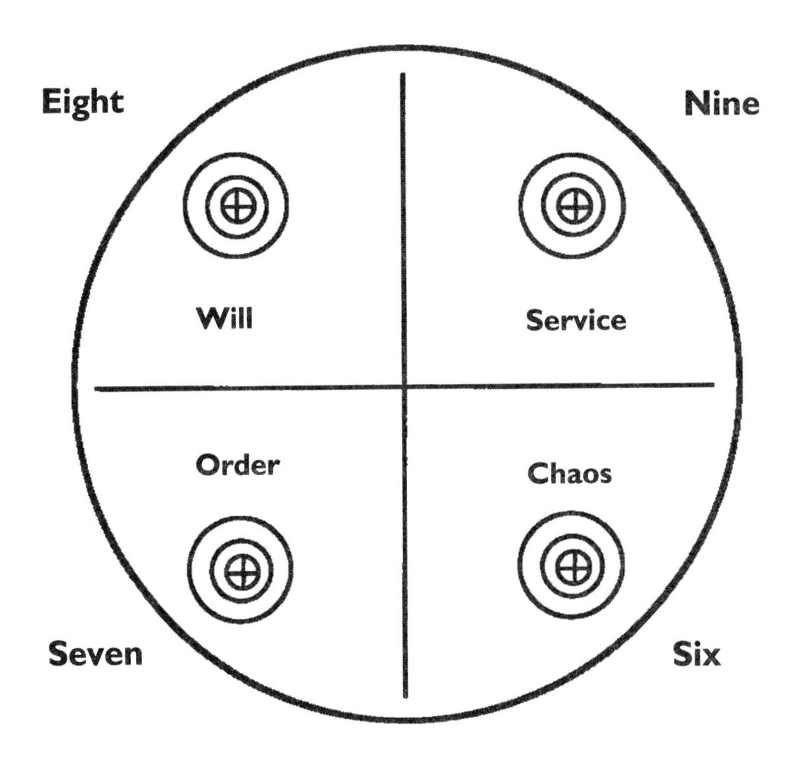

The Chaos game
converts mastery of Will
and observance of Order
into the act of Service.

Vacuuming the Shadow

Low into the mud I crushed myself, pushing down and away from the heat scorching my back. The fragrance of sizzled herbs filled my nostrils and lungs. The darkness of the sweat lodge notwithstanding, I could see Grandfather Keetoowah, shriveled in his illness, a few feet away. The intense heat squashed me down in search of muddy relief but did not faze the old Cherokee. The sweat leader prayed and ladled water over his crippled shoulders, and clouds of steam hissed as Keetoowah groaned and shouted. Pain evolved into time and returned as shadow power rippling through his body.

Keetoowah, who always said he lived in the alpha state, was dying, but the ceremony restored him for this autumn night. An apprentice wrapped a blanket around the tiny man and carried him like a child from the lodge.

"It was a good lodge," Keetoowah said quietly to the apprentice. "I had to get that energy off me. Been on me for a long time, for years, on my back. Damnedest thing I ever seen. I'd have died tonight if I hadn't got it off me. It was a good lodge, hot enough to get rid of it. I feel better now. Worst thing you can imagine." The apprentice gently placed Keetoowah in a porch chair and bundled him up.

"Grandfather is very ill," confided the sweat leader. "But it was a good sweat. I saw a heavy burden pass from him."

"Really? What did you see?" asked a guest unfamiliar with sweat lodges and their language and etiquette.

The leader looked away silently. Finally, he just said, "Whatever it was." Then he returned to the lodge to lead another sweat session.

Grandfather's etheric burden rose and merged into the cosmos while the sweat lodge shimmered in ceremony under the full moon. I crouched on a hill of pines above the lodge to watch attendants carry in hot rocks and listen to the outflow of muted chants and drumming. After a time, the flap opened, people crept out, dowsed themselves with cold water, stretched out to rest on Mother Earth, and one by one disappeared into nearby shelters and tents. Silence assembled its night. A screech owl shrieked, then fled its pine on silent wings. The south wind arrived amidst wisps of spirit clouds. Sleep did not tempt me as I leaned against a fragrant pine. The trance arose easily.

"There are two aspects of knowledge," Wolf said. I had heard the teaching many times before. "The first is self-knowledge and the second is magic. Learning magic without first mastering your self opens one to the influences of evil. All forces of the universe are potentially very powerful. If you have not cleansed the self with understanding and release before accessing these forces, then possessive energies can inflate obsessive fantasies and hidden agendas into incredible and even deadly proportions."

Photographic images of massive death scenes at Jonestown flashed before me, as tragic as any war. Everyone wants the magic, the power. Few are willing to work for self-knowledge. The result is obsession and, taken one step further, possession. If the human shadow is not cleansed of the fear of unknown darkness, death becomes the manifestation of that fantasy, a combination of terror and oblivion.

"Let's play the mystery game," Wolf said. "Breathe quietly and concentrate on recalling your vision of kaleidoscopia, the world of the grand tumbler and the four smaller tumblers of hidden agendas. Remember that chaos is northeast, order is southeast, will is southwest, and service is northwest. Also recall your understanding that these elements are problems because they are hidden agendas. In other words, you maintain these concerns in the back of your mind at the same time you attempt to deal directly with the phenomenal world around you. This means that any attempt by you to directly perceive the world in front of your eyes will be complicated by your inner need to first interpret that perception in terms of your hidden agendas. You may suffer from all four, you know. Walking on both fantasy road and coyote road is a complex exercise in awakening awareness.

"There is one way to permanently escape this agenda trap," Wolf went on. "You have the power to convert the four agendas into four more guiding spirits of Twenty Count. These spirits are named, of course, Chaos, Order, Will, and Service. Once you embody them as aspects of yourself, you will stop being the victim of your underlying need to manifest them. Your shadow will be cleansed and will assume its new responsibility of transforming former fantasies into present possibilities. Think of it this way: the chaos game converts mastery of will and observance of order into the act of service. The base of this act is the capacity to isolate your identity through perception without alienating your self through reflection. The result is a balanced projection of who you are.

"This is another way of describing the art of passionate detachment. No matter what we call it, this process induces a deep, inner feeling of compassion. Don't let similar terms confuse you. Passion is your own fire, your intensity of motivation. Compassion is the all-embracing warmth of the ancient heart, the fire within the primordial flash that enlivens you and all other beings.

"The act of service, you see, is the embodiment of passionate detachment, a state available to only those to whom fantasies do not appear and are therefore unaffected by hidden agendas, those unburdened by ego demands. To practice passionate detachment is to express compassion through service. You need not establish service as your goal. It's something that just happens during your awakening. You will discover your own act of service, straight ahead, and it will amaze you with its simplicity."

Wolf instructed me to visualize the grand tumbler of kaleidoscopia as a planet making a quarter-turn on its axis. Each of the four smaller tumblers within would appear as continents moving clockwise and assuming new positions. This is the process of shifting personal perspective in order to convert hidden agendas into spirits. With the quarter-turn, Chaos is now southeast, Order is southwest, Will is northwest, and Service is northeast. In this new world order, Service emerges as the guide of the location for making choices and moving into freedom.

The four new spirits, with their unique history as former hidden agendas, assume powerful new roles as cornerstones of identity. The pathway through these cornerstones into appropriate actions and relationships bypasses the tiresome fantasy road and reveals a clear view of the self's valid identity as observer.

"The method of cleansing through understanding and release is called vacuuming the shadow," Wolf said. "Only when the shadow is thoroughly cleansed do you have inner space to store up the power necessary to instantaneously overcome every trance, spell, and curse that you will encounter on fantasy road and coyote road. You started your journey by asking for clarity. Clear sight is available through the vacuuming process of regression and progression. Back and forth your perception and reflection travel through Twenty Count until you establish a sustainable vision of the ancient heart. Then you can realize your true identity by blending chaos and order into will for service."

Wolf's explanation soared through my mind. I had the feeling of staring straight into an immense central realization while lacking the power to prevent it from both eluding and befuddling me at the same time.

"Progression and regression," Wolf explained more slowly, "is your path of moving through the Count. Your breathing, in and out, out and in. Your existence, life and death, death and life. Your thinking, past and future, future and past. Your perception of phenomena, chronological and synchronous, synchronous and chronological. Your essence, here and there, there and here.

"Clarity is the condition you attain when all these seemingly opposing conditions are encompassed by expanded awareness, and you experience absolutely no need to interpret them by examining their apparently contradictory natures. You feel no need to complain or to think of yourself as being at the mercy of their potentially chaotic power. As spirits, Order and Chaos balance one another, permitting Will to emerge as Service. Enlivened spirits are powerful enough to choreograph dissolution of fantasies from which their original roots sprouted.

"When you achieve this balance, whether through breathing or meditating on the Aztec Sunstone or grinding your intellect through mythological thought patterns, you will have vacuumed your shadow. The art of passionate detachment is merely a tool that will help you eliminate the need to fantasize, a sort of power cord for your vacuum. The act of service becomes your unavoidable next option. This act infuses itself through your entire being and commences its process before you can even acknowledge its presence.

"The vacuuming removes the blackness from the diamond that represents self's reaching out, and it creates space for Child's cross to ground itself within the circle. If you recall the coyote road condition of pain resulting from the cross with no circle, you will understand how releasing the diamond's blackness leads to relinquishing pain. This is also the lesson of Elder to Child: to embody teachings within the body. In essence, to rediscover your original identity as observer, to understand that you are a dynamic being. You are the very light emerging from the ancient heart. Perception of your existence does not require limitations or solid boundaries in order to retain identity. You can exist as separated self without being dominated by the fantasy of alienation. If clarity is your true goal, the only required capacity is to see."

Chaos, Order, Will, and Service manifested as four corners within my wheel of vision. Unlike other spirits, they did not speak to me directly, but became constant reference points for my familiar spirit speakers, Wolf and Old Ghost. Holygram also hung around, offering his colored lenses through which I viewed the vast realm of kaleidoscopia and the interaction of the four corners. In time, the four merged into a collaborative unit whose purpose evolved into coordination of the quest itself.

Acceptance of the noncontradictory nature of the four spirits was essential to vacuuming the blackness from the human diamond so that the technicolor world of kaleidoscopia could emerge. When the blackness disappeared from the human diamond, it also vanished from the smaller diamond configurations of the tumblers. When only whirling brightness remained within all aspects of the grand tumbler, the hidden agendas had vanished. I accepted clarity in my life by assuming my original identity as observer. I also commenced a new vigil, this time watching for my act of service.

"Quite a world you've assembled here," Old Ghost said one day. "You do realize, don't you, that you've been vacuuming the shadow since way back on Algonquin Peak? It's all step by step. You spent years accumulating information by studying religious and cultural forms. Next, you converted information into knowledge by embodying the contradictions of Chaos and Order as spirits. Now you can activate your Will to transform knowledge into wisdom through Service. After all this, do you think you're ready to find out what your big act is? Are you ready for your mission?"

"If you say so, then I'm ready," I answered. Long ago I learned not to argue with Old Ghost. Or any of the spirits, for that matter. Yes always kept energy and information flowing, especially whenever I wanted to say no.

"Then be on the alert," Old Ghost replied. "At the right moment, your work will be revealed. In the meantime, just know you're already performing it. Don't let chronology confuse you. Remember synchronicity. Whatever is happening right now also goes on in the past and future. When you are ready to reenter ancient heart, your act will become so obvious you'll wonder why you haven't remembered it all along. You will realize just how much you've forgotten."

So I began the wait for my true work on Earth while attempting to manipulate my journey through Old Ghost's linguistics of time and space. Vacuuming the shadow involved integrating various components of therapies, philosophies, disciplines, and creativity into a clarified focus on releasing emotional bonds that blocked or otherwise disabled clear-minded and heartfelt thought. Clarity, my purpose for years, transformed into a discipline of active passivity. It demanded that I remember my forgetting. Otherwise, I would reach neither understanding nor release.

Clarity stressed the premise that no matter what religion or belief system one subscribes to, understanding and release are possible. Religious fundamentalists may ask forgiveness of a deity through their worship process. Psychology subjects probe their dreams and childhoods. Reincarnation subscribers regress many lifetimes to unearth roots of problems and solutions. Perceivers of concentric existences travel to the realm next door, or down the astral street.

Regression processes may involve opening a series of ancient eyes and ears, or recalling past lives by cracking the occult code of mental obscuration that blurs deep memory. One teacher showed me a complex but powerful meditation to sever karmic bondages by releasing a flow of medium-blue energy from the medulla oblongata at the base of the brain through the jaw hinges to ancient eyes in the palm of the right hand and sole of the left foot, then back up to an ancient ear at the crown of the head.

On the other hand, a good belly laugh at one's own fantasies might be just as effective. However it is approached, the vacuuming process involves unraveling your life tapestry and reweaving it into a restructured pattern.

All these considerations hovered into mystery game revelations that night, as Wolf and I kept vigilance above the sweat lodge and the sleeping Crystal Godfather. Midnight moved toward dawn as the trance shifted and a new speaker surfaced.

Keetoowah appeared imminently familiar, still an aged elder but much stronger and healthier. He wore his white turban, the formal Cherokee headdress, and sat behind a table, a huge crystal ball before him. Though clear, the carved quartz revealed a second round crystalline image within its form–a ghost of its own being. Keetoowah often spoke of being consulted by other healers on crystal matters, and of being the only man allowed to cleanse a particular crystal found on the Atlantic Ocean floor and linked to the ancient culture of Atlantis. Now Grandfather gazed into what surely must be that crystal.

"Crystals are incredibly powerful," he said. "You can see anything. I look into this stone and see another crystal within it, a joint double with a magic whirlwind at its center. And whatever you can see, you can wish for. But you have to be careful, very careful, about what you ask for. If you ask for something material on the material plane, you'd better be real clear or you'll end up paying for it in ways you can't imagine. Crystals are great generators. You can use them to generate the energy to create whatever you can imagine. So you'd better have a real grounded imagination when you use them. The Cherokee train their healers for years and years before they're ready to use the medicine. You've got to get your mind clear. Then the magic that can come is incredible. That's what happened to me. I went through years of experience, all kinds of diseases, all kinds of healings on myself and on others. Now I live in alpha state, and I see and feel the magic everywhere."

"Grandfather," I said. "Your back is healed, is it not? You look so strong. You seem clear."

"My double has never been pained. It has guided me and I have finished my service," Keetoowah said. "I have taught you, the children. Now all my pain has gone. The sweat took it away, and I feel the great whirlwind directly. When you think of me, you must remember me as healed, not hurting. That is my true identity. When you know your true identity, you can endure your own whirlwind."

He looked up from the crystal and straight at me. His eyes were moist, and I flashed back to our first meeting. The crystal I had used as centerpiece for my arrowheads meditation was a tiny skull carved by a Tibetan shaman. When we first met, I showed it to Keetoowah. He had taken the crystal gently and held it in his hand. Tears welled up in his eyes.

"Some of these carved skulls hold real harsh energies," he said. "They have been used for all kind of evil. But not this little guy. He's good. You know, I've never carved a crystal skull. But I like this guy. I'm going to make one."

He later found a crystal chunk from which he planned to carve the skull, and he eventually started the slow, painstaking project. But crystal carving is demanding work even for agile hands, and the aging artisan was slow to the task. Years passed. To the best of my knowledge, he did not complete the carving.

Late in 1987, a few months after the healing sweat, I learned of Keetoowah's passing. At midnight, I walked out to the end of the long pier at Avila Beach near San Luis Obispo, beyond the lights of the little town and beyond the night fishermen. A large seabird suspended active flight to float on air currents overhead. I gazed as far as possible into the black sea.

I grasped the tiny crystal skull as Keetoowah himself had done, held it out, and dropped it into the ocean. A ring rippled outward from the spot where the skull disappeared, then a second ring, a third, and many more, radiating like a little oceanic whirlwind. How far the ripples would go, no one could know, but my focus fastened intently on the skull's point of disappearance. The ancient heart of my mystery wheel was fixed forever.

Good night, Grandfather. I will remember that your pain has gone.

Good night.

WHIRLWIND MAGIC

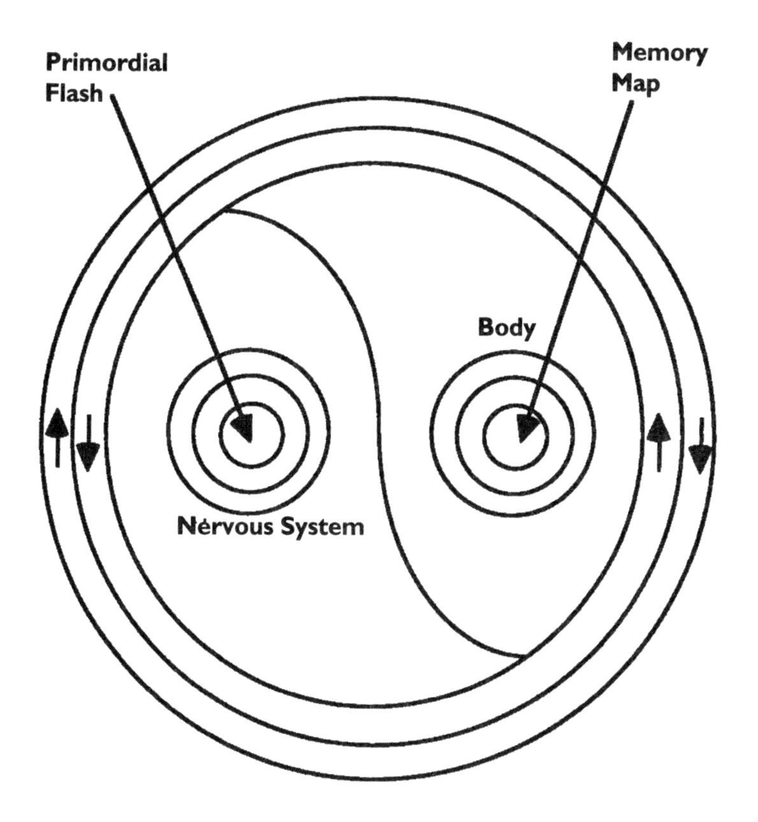

Out of the flash
into the mind.
Two flowing into one.

Joint Double

Delicate chips of memory haunt this body I walk in. Spasms of shadow-laden light cramp the lower back and soles of the feet. Fever and paper cuts betray meditation. In the war to return home, the flesh becomes the casualty. My nervous system is the history of love and hate, and it lusts for command. When I sleep, separate worlds demand my attention to their unassimilated forms and content. I can see them all, I can hear them all. If I deny this range and potential of awareness, a billion secrets will vanish undeciphered when the body dies. The only chance to access the hidden knowledge within the muscles and bones and tendons is to discover my inner code of knowing during this jaunt around Mother Earth.

Waking midst sweat and chills, I opened my eyes and called back the realization–the code. What was this code? A dreamlike vision of flashing feline eyes flared and blurred and disappeared.

"My code," I called out as I sat upright in bed. "My code, my code, my code."

"Your code, your code," a laughing woman's voice echoed from far away, "is the joint double. When you master the joint double, you will remember it."

"Joint double? Joint double?" The words rang true, and I wrote them on the note pad by the bed. Giving way to the fever's drain, I lay back down. "Who are you? What do you mean? What's the joint double?"

"I am Dream Healer, your friend. Remember me? Good, that's a start. The joint double is your key to making sense of the wheels and stories and visions and voices you've been gathering together for years. In your fever, you are dreaming. In your dream, I am healing. When you wake, the fever will be gone. In its place, you will be aware of the joint double as your key for reading your memory's map of hidden secrets."

Dream Healer was the first definitely female spirit who had appeared from Twenty Count, and I had yet to establish a clear image of her. But now a beautiful naked woman formed to my vision's delight, tall and strong and peering straight into me. Her eyes and hair and skin slowly changed hues, ranging from dark to light to bronze. A different heat supplemented the fever. My throat was dry and tight. I wanted water. Yet nothing existed powerful enough to distract my observing. The pain of the heat thrilled me. I felt ill—or perhaps a bit insane, whichever condition filled the vacancy of my old mythology's unconditional defeat at the flashing vision of ancient heart.

"Just as personality must submit to its baby death," Dream Healer said, "so must the physical body release its essential secrets through the joint double. In accordance with the orderly reflection of existence, your body is a double of your mind. Every secret lodged within the mind is perfectly reflected somewhere within the physical being. Sensation in the body is an opportunity to contact the creative center, but usually the mind can't handle direct perception of pain. Think of it this way. The primordial flash is the heart of the nervous system, and the body manifests itself around the mind's memory map, the secret-laden configuration of the original transmission of existence from source in the heart to sensation in the mind.

"As you travel the spiritual path, you learn to consider personality as the arch enemy of inner growth and the body as an impediment to spiritual awareness. In truth, both personality and body are highly trained agents of the spirit, and both have essential missions to accomplish in terms of your capacity to awaken to the incredible essence of your dream world, your holographic kaleidoscopia. Personality is committed to exposing the utter futility of seeking inner meaning from outer sources. When this futility is realized and embodied, observer easily transcends baby death and turns a stronger and more committed attention toward silent instinct.

"Now it's the body's turn to discover its mission. That's what you're doing right now with your little fever and shivers. The concept of joint double has been your personal mystery game and mind barrier since you glimpsed your first sight in this life, since you heard the first utterances that landed you on this path of exploration. It's been One and Two from the start. Perception and reflection gave birth to fantasy. Remember? Chaos and order. Will and service. Back and forth. The duality that turned into projection and raised the question whether everything on Earth is in conflict or cooperation, and in turn brought you to the point of choosing whether your life is full of discouraging blind alleys or wonderful hidden passageways.

"The joint double determines your perception of lunar and solar, male and female, child and elder, the very cornerstones of identity that result from vacuuming the shadow. When your vacuuming is complete, whirlwind magic emerges as an incredible and explosive union of magic mind and wind power joined in the process of aligning phenomena into essence. The joint double whirling inward toward its own ancient heart is human sexuality. Its incredible yet elusive potential is the body's most obvious secret: plain old orgasm."

Dream Healer must have been reading my dream mind as I kept my inner eye glued to the beautiful naked woman of fever and chills. A delicious, driving lustiness was building within my heated physicality, even as I focused my hearing on the ethereal lessons of Twenty Count's spirit of the moment. Dream Healer, I thought as vividly as possible, is this you I'm seeing?

Dream Healer's wonderful laugh burst through me just as the beautiful woman transformed into the physical image of the perfect male counterpart to the woman of my fever. He was every bit as striking as she. Then fever woman returned, then the male resurfaced. Somewhere in my physical body, the fever broke.

"Allow me to introduce the only Twenty Count spirits you have not yet met," Dream Healer said. "Your own joint double within, Light Man and Fever Woman. Maybe they will help you remember how much excitement exists within your own body, your own sparkling soul, completely separate from any other physical presence. These are the spirits of Eleven and Twelve, raised inspiration and self-based order. They have always been part of you and always will be. But take my advice and don't get hung up on listening for their voices. I talk to you, and Wolf and Old Ghost talk to you, but let Light Man and Fever Woman simply manifest their presences by balancing your life. Just as Order, Chaos, Will, and Service have become part of you, allow both female and male to simply become partners within your silent instinct."

Dream Healer's suggestion inspired a revisiting of the innermost realm of primordial flash, the core ring revelation of lunar-solar insight. Light Man and Fever Woman leaped from the depths on the back of Fox, soaring fearlessly into such direct perception that hidden agendas had no opportunity to arise.

I sensed Fever Woman proclaiming: "I am the order running through all experience, the order without which you cannot structure realities." And Light Man saying: "I am the wind, the movement you need to perpetuate perception." Together, they thought- sang in the tradition of Child and Elder: "We create whirlwind magic. Our possibilities, which seemed limitless yet strangely inaccessible alone, now embrace new potential for reality. Together, we feel the inner fire and hear the music playing. We explode in perpetual orgasm, energy burning us into an explosion that spews our secrets out into all-encompassing vibration. Uncontrolled elation erupts through our being like primitive lightning as validation of our instinctual wisdom concerning our source, existence, destiny, and drive to perpetuate existence. For a fleeting moment of ecstasy, light is everything, the apex of our thrusting and thirsting. Flash yields sensation, perception inhales light, life imprints memory. We scream and spasm in the ancient heart as though our bodies were boiling in pain."

In the passing moment, though, the ecstasy abates. Aloneness replaces the spasming flash. The silent instinct of perpetual wisdom, knowledge, and information recedes like a tiny dot on the mental horizon. Fever Woman wants to hold the vision, but Fox is ready to fly. Light Man consoles her: "We will do it again." Fever Woman leaps for Fox, catches a ride and flies on alone. "Yes," she shouts back to Light Man, "we will do it again."

Like Fever Woman and Light Man, we humans play our scene over and over until we can fit the cornerstones of identity into the joint double's magical whirlwind. Ever-present ecstasy hides deep within us until we seek it out, and then we grasp it for only a moment. We let it go, remembering its thrilling form while forgetting its boundless content. But we try again, seeking the thrill anew. We leave ourselves clues about the possibilities of deep memory within the process: feelings, shouts, surging heat, and tingling shivers. These hints inspire us back into the process, time and again. We seclude the clues as secrets in the folds of magic mind. And we sense, through silent instinct, that we can access our memory map's guidance through the presence of a joint double.

"When your joint double emerges, you get a feeling of becoming complete," Dream Healer explained. "Just as you feel better now that your fever has passed, and as you feel more solid now that you have met all Twenty Spirits. That feeling you're experiencing is coming from holding your focus on the creative center long enough to get a good look at it. Then you methodically step up, look over the edge, with or without Fox's assistance, and take in one good perceptive gulp of the ancient heart. This means you have accessed your inner code and are ready to remember."

With the joint double present, I did remember, at least to some degree. Scanning my remembrances as the sum total of many different aspects, I was taken aback when one inner entity addressed me directly: "I am the shadow entity. Vacuuming has transformed me into the illumined shadow, but my development is incomplete. If you educate me and establish my identity, I will become reality. I yearn for completion through sexual union, just as you do. I must take on identity, then I can join you to create magic in your material world.

"As Freud might have said, an identity without an id is only an entity."

"What?" I awoke, still groggy from the fever. Somebody was summoning me to consciousness with a bad pun.

"Welcome back to alpha," said Wolf. "Now that you've located your identity and your joint double, maybe we can get on with some serious remembering. Just relax, you don't need to drink yet. Use your thirst as a concentration point for forging into remembering. See yourself as being very young, just a little tot. Forget your age, your history. That's good, good. Now what do you remember? What's happening?"

The universe swirled as a mass of liquid energies flowing into and around and through one another, a whirlpool of electric colors. From the whirlpool emerged the physicality that would embody my aspect called Child. The original choice to be male or female had been so effortless that Child did not recall the choosing. Rather, Child remembered only the experience of inhabiting a body of the half-self, a separated physicality encasing a yearning to touch and be touched by other physicalities. The touching brought back the feeling of the flash when all was One and unlimited possibilities flourished. Child grew in the body of half-self and, with Elder's teaching, learned the Earth role of man.

"As you remember your spirits as both Child and Elder," Wolf suggested, "let Child take the lead this time. Let Child tell Elder what is most important to be remembered. After all, Child was over in the hither world much more recently than Elder. If you can access Child's magic mind through remembering, you do not need to rely on Elder's telling of myths and teaching as your source of information and guidance toward knowledge."

Child's words rang melodically: "Please, Elder, I know you want to speak to me. But first, please listen to my story of travel. Listen to what I ask. You and I can give so very much to one another, but you must listen as well as speak. I have so much to tell you that will help you teach me. I have just emerged from the hither world. You know, the other side of the ancient heart. I know what I must learn in order to grow, if only you can learn to listen to me before you speak. In this way you and I are the joint double. I know it is difficult for you to hear me, but I must tell you what I need to learn.

"Teach me first, then tell me later. I must be grounded before I can be magic. Teach me that Order is a powerful guide when heeded, but a deadly enemy as a hidden agenda. If I can comprehend and apply this teaching, I will be prepared to hear your telling about the power of Chaos, Will, and Service.

"But before and beyond all, teach me that I can live free. There are pleasures and pains in your mythology, and you think all you can offer me is your knowledge. But also teach me not to hook my identity so deeply into you that I think I am you. I am not you. Let me see that I am separate from you in thought and action, while a part of you in spirit. Do not fear for me as you prepare me for the world. If you teach me well, I will survive. Help me understand that the universe issues an instantaneous challenge to our every demonstrated weakness, and that the purpose of that challenge is to create strength.

"Open the sexual world of duality for me. Love me, but do not make me dependent on love, neither parental nor sexual. Make me aware of my own generation of love within, its power, and my capacity to center the teachings within myself so that freedom emerges. Do not try to own me. Children are not possessions. Do not live vicariously through me. You must nurture me, not I you. When I am adult, you and I will treat each other as adults. When you are old, I will care for you as a matter of the flow. But still, do not expect it. Do not rely on me, for then I must teach my own children, another generation coming after me. Life must move forward, not backward. By the time I grow into Elder, I must be strong and powerful, not weak and pitiful. Teaching me this way gives me a head start to freedom. As adult, I will remember my lessons of Twenty Count."

"Child is wise," Wolf confirmed. "If you can remember Child's plea, you will understand how to develop wisdom that opens you to remember and apply passionate detachment, the art which illumines death as a part of life. But you first must complete your spiritual journey. You've been searching for years, and you've discovered an identity. But now, you must go for the magic, the reality of the whirlwind. You must learn to manifest substantial results in your material world. And you want to know something? Your joint double is also your code for success. It's still your key. Yep, your joint double is quite a cat. Quite a code.

"The reason you need the joint double so much," Wolf continued, "is that all this wandering around in kaleidoscopia doesn't create a grounding effect. But when you access appropriate magic by absorbing and applying the wisdom of both teachings and tellings, you will activate your very own thought atom of alpha empowerment."

The thought atom of alpha empowerment. Whatever the phrase meant, it rang powerfully clear. Accessing my joint double code, I passionately detached myself and steered as many of my inner conflicts into resolution as possible. In time, my joint double manifested as a woman whose perceptions and expression aligned in clear complementarity with mine. She and I learned to speak about the teachings and tellings in a way that reflected the raised inspiration and self-based order that twenty spirits had manipulated into my memory map. One evening as we mused over Twenty Count, I attempted to explain the joint double itself. I wasn't doing well.

"It's a matter of being clear," I said. "We all need to perceive as lucidly as possible what's happening in every moment. And ultimately, that means acknowledging that life is a process of dying, and to pretend otherwise is to be in denial. But to maintain that focus continuously is virtually impossible. I mean, we all need to sleep sometime. As the philosopher Ouspensky taught and as Aldous Huxley said many times, our business as human beings on this planet is to wake up. So we need our other sides to keep snapping us into the awareness of the intensity and potential of life or we'll never be able to experience and embody that intensity and potential. We'll be stuck forever floating in fantasies."

"I've always been afraid of death," Haleh said. "The idea just makes me tense. It makes me block things out with a feeling like panic. If you and I are creating a joint double, it seems that I need to resolve this fear, right? But death is something we all must go through alone. What good is a double at the moment of dying? What possible purpose can it serve? What is the Twenty Count basis of resolving death through the joint double?"

"It's about the return," I answered. "It's about silent instinct. There's just something we know somehow about the other side. The other side of ourselves, the other side of existence, the other side of Twenty Count. There's a sort of natural nobility in recognizing and observing the order that creates unity within a joint double."

"Natural nobility?" she responded. "You know, I do feel that. I even almost understand logically what you're talking about. Can you say it another way?"

I began humming only slightly off-key as I awaited renewed inspiration. When my mind needs relaxing, I recite or sing little chunks of familiar poems and songs, often hits from the 1960s. Certain passages almost seem coded. This time, however, the song was more recent. Pondering Haleh's question, a chilling passage from Leonard Cohen's song "The Captain" leaped from the map of memory:

> There is no decent place to stand
> in a massacre.
> But if a woman take your hand,
> then go and stand with her.

"Yes," she said. "I get it. Now, please don't say anything else so I can just hold this, and don't ask me to explain right now. I don't want to lose this. But I get it. Somehow I get it."

This elusive, enlightening wisp of comprehension that we get but cannot explain is the thought atom of alpha empowerment, the power of Eighteen.

THOUGHT ATOM

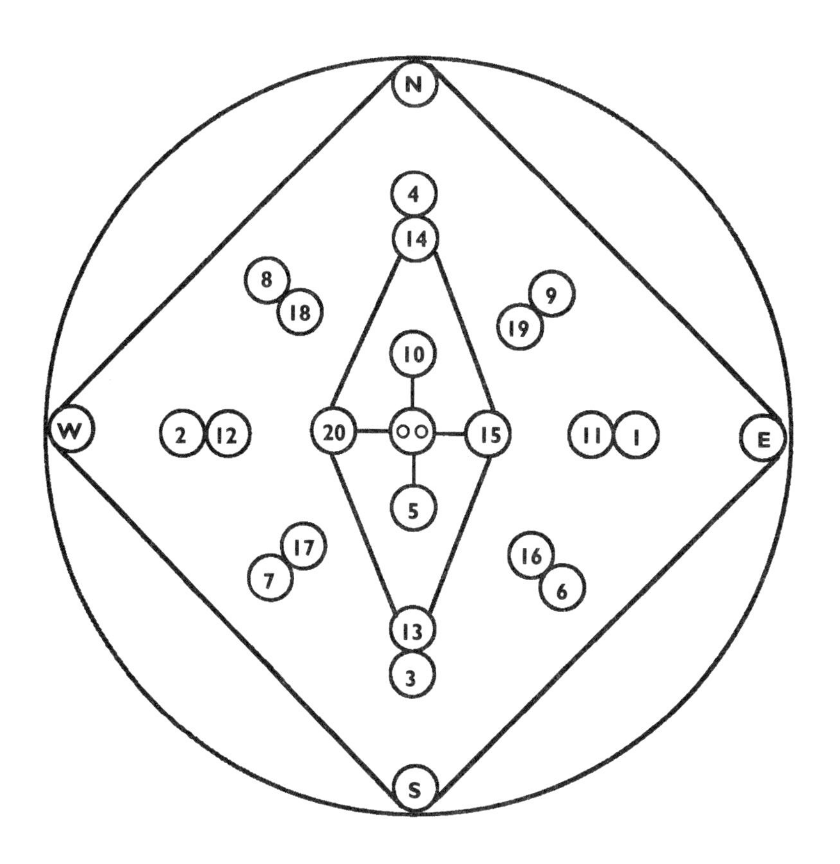

Seeding cosmology
in the blink of an eye.

Stew in the Matrix

Leslie Gray is the only shaman with an office in sight of the high rises of San Francisco's financial district. An Oneida healer with a Ph.D. in psychology, she practices the teachings within a modern-city maze. A vibrant woman, Leslie incorporates techniques of ecstasy from throughout the world into her work. Appropriately, her business logo is a vibrating four-directional spiral.

A dozen people sat circled on the floor in cross-legged poses as Leslie led the drum and rattle ceremony. Amid the pounding and rattling, a shrill but seemingly far distant trilling rang out and sent chills through my body. I could not count how many such circles I'd sat in over the years, but nothing had jerked my insides like that high-pitched trill. It came again. I resisted the urge to open my eyes to determine whether the voice had risen from someone within the room, and to this day I don't know. The rhythm continued, and my sinuses responded by opening as never before. Breathing deepened and became effortless. When the drumming ceased, everyone lay flat on their backs while Leslie's voice guided the journey into both lower and upper worlds.

Until this night, my most vivid experience through shamanic journeying had been in Scotland where the people also work with feathers and crystals and chants. The leaders had summoned up magical beings such as nature spirits and devas who brought messages about the need to reconnect our human roots to Mother Earth. But something more tangible was in the air this night. I'd been in California for a few years, putting together images, interviews, and information for a book on the pursuit of awareness. Now the book was near completion, but I had felt powerfully drawn to do one more interview, specifically with Leslie Gray. Graciously, she had invited me to participate in this journey.

My inner view of descent into the lower world was like climbing down the walls of the Grand Canyon just at the break between sunset and twilight. A cavernous depth emerged, bathed in a mixture of fading reddish light and deepening shadow. I sat on a rock and waited. Without thinking, I posed a question aloud:

"What's wrong with the book?"

An old woman came from the shadows, and my mind shut down. All I could remember when I returned to Leslie's office floor was that I'd received a surprising and rather unwelcome bit of guidance, which I later followed and which altered my path significantly. But whatever had passed between me and the grandmother spirit after my surprise question and her surprise answer was vague and blocked out. By my returning again and again to the vision and, with Wolf's gentle prodding of my deep memory, the sequence of our communication eventually surfaced in consciousness.

"She was your Karma Master," Wolf explained. "Your spirit

for Eighteen. Her purpose was to put you into as direct contact as possible with One and Two, the spirits of Serpent and Mystery. She accomplished this by explaining how to merge the identity you established at Sixteen with your magic at Seventeen in order to plop your joint double into this bubbling stew in your matrix at Eighteen. In other words, she guided you directly into alpha empowerment."

"Believe it or not, I almost understand what you're saying," I said. I had never actually seen Eighteen clearly until now. Apparently there had been too many ties that blind. "Please go on."

"Karma Master is your stew guide, but Lama from Sixteen, the place where you resolved history, is the cook whose recipe creates the stew. Once you learn to nourish yourself with the stew, you will enter what Keetoowah called 'living in the alpha state.' In this state, you can ascertain the hither world. But more about that later. Let's take it step-by-step for a while."

Wolf's step-by-step went like this: The spirits of One and Two are Serpent and Mystery. Serpent is the first entity perceived by the observer. But because no reference point exists for identifying Serpent, the observer must rely on instantaneous interpretation to gain an understanding of this writhing, multicolored energy. By the time the observer goes through the perception-reflection-projection-balance process and comes up with a story about the object of seeing, the original entity's actual essence is long forgotten by the observer, its impression logged somewhere deep in the folds of memory. Getting further lost in the workings of the process, the observer identifies with the observation and constructs a belief system that relies on the limitations and interpretations of this process, which is full of misdirection because it lacked reference points at its inception. The observer's clouded identity process thus becomes the basis for separated self.

The observer goes on interpreting sensations by realizing aspects of separated self as history, dreams, relationships, and choices. The existence of inner awareness is determined, and this leads to renewed inspiration, a sense of orderly organization, a transformation into living the inner life, and an onset of instinctual awareness. That's the process from One through Fourteen. When the observer hits kaleidoscopia at Fifteen, the holographic scene presents a telescoped universe of thought extensions shooting out from every original interpretation as perfectly aligned fantasies and possibilities.

The observer's new open-mind policy permits a release of ties to history at Sixteen, where fresh cornerstones of identity have been established. These cornerstones create a realization of self as a being embodying both Chaos and Order while also possessing both individual Will and the desire for Service to others. Once grounded by this abstract realization, you–as the observer–are ready to master the magic of Seventeen by reframing reality in terms of a synchronicity/chronology synthesis. This process leads to a more direct understanding of Serpent, the spirit of One.

"This process also reveals the true nature of our nagging internal mystery," Wolf explained. "Mystery is Two, the reflection of the Serpent, the spiraling nature of life in all its manifestations. It's in every being. It sits in the human at the base of the spine, and it lies buried in tiny sparks of secret stuff throughout the mental and physical aspects of the entire body. But you have a terribly difficult time just getting your tools of perception sharpened to the point where you can capture this reflection of Serpent. Beyond that, your interpretations of Serpent as an embodiment of evil have convinced you that it's a scary, terrifying beast that can only hurt you. I'm rather sensitive about that because you've also pinned that stereotype on me. Your kind is scared of Wolf, my kind. All because you can't remember what you're looking at. You just can't remember that Serpent is a spirit. You can't remember that I'm your soul."

Accepting Wolf's premise that we humans are bigoted against serpents and wolves, I apologized profusely. Wolf moralized that it was about time and then ushered me further into the explanation of the stew in the matrix at Eighteen.

"The Lama who released you from the binds of history by digging up the seeds of blame and praise knows how to prepare your stew, blending in the zestiest ingredients from your place of all phenomena and seasoning the whole concoction with lessons learned during your quest. The stew then becomes your nourishment, the blend of your life experiences, your renewed sense of who you are, and your magical capacity for seeing life directly and without a bunch of phony personality cover-ups. What comes out of all this is a special being to guide you directly to Nineteen, the gateway to freedom, and then to the ancient heart at Twenty."

According to Wolf, the power of the stew in the matrix was a special spirit, more unusual than even the surface spirits of Twenty Count. That spirit was Hummingbird, the beautiful, darting, dashing little flyer with the long beak, neon breast, and incredible ability to hover wherever with wings beating at invisible speeds.

"But I thought Karma Master was the spirit of Eighteen," I said.

"She is," Wolf agreed. "But Hummingbird is her friend, her familiar being, you might say. Hummingbird pops out of the stew to carry you on over to the hither world, which, by the way, is what you refer to as the other side of Twenty Count. That's where you end up when you go through the center. Surprise, surprise, there is something there. You just have to remember what it is."

Hummingbird's name was Alpha. She was the most unique being I had encountered in or out of dream worlds. Passionate detachment was easily observed in her every fluttering moment, and she had no time for hidden agendas. She sped out of sight in the blink of an eye, but always returned when I needed her wisdom and viewpoint. Talking to her was a waste of time, and I stopped trying. Finally, I became accustomed to her presence, even when she was absent.

"She is part of you," Karma Master explained. "But she is even more a part of me. I know her better. And I know what you want to remember."

Sitting on my lower-world rock, zoned out in a shaman's downtown office, I felt the question return: What's wrong with the book?

"It's very simple," Karma Master explained. "It lacks the necessary thought atom."

Every time I felt close to comprehending the Twenty Spirits and their explanations, they had the wonderful technique of introducing a strange new concept that usually produced a series of reactions. I often felt frustrated, elated, and manipulated. But I also felt honored because I had learned their concepts always made sense.

"I will explain," Karma Master continued. Her voice turned kindly now, motherly. I felt safe and at home in her presence. I sensed the power of being beyond harm from the phenomenal world.

"One way of examining the structure of your thought is through the possibilities and probabilities of a thought atom. Each individual thought atom, including the initial realization of the perception of light, is constructed from some combination of Twenty Count's strengths, weaknesses, and openings. When you look at Twenty Count, in any of its many forms, you are seeing a thought atom. Concentrated focus will reveal within this atom the process for developing a sophisticated concept.

"As a human, if you wish to communicate meaningfully with another person, you must transmit a completed thought atom to that person's perception. You must provide that person with reference points so that perception can be grounded in a common reality. You must consider and eliminate tangents and possible pathways of alternative interpretations so that fantasy does not take over, and you must demonstrate through communication how the completed thought relates to your relationship. Otherwise, you're babbling. Of course, babbling makes up a tremendous amount of the so-called communication that goes on between humans. But if you are working with the twenty spirits, it's really quite unacceptable."

"You just told me that the book I'm writing lacks a thought atom," I said. "In other words, my communication on behalf of the spirits is inadequate."

"That's it," said kindly, motherly Karma Master. "Stop the book. Destroy it. Start over. When you've mastered a thought atom, you'll know it as a self-evident truth. It will feel like a code revealing itself to you."

"How was your trip?" Leslie asked after the group had awakened from its drummed-up dreams.

"I got a message. I don't remember all of it, but I sure remember part. It was very clear. An old woman came and spoke to me. I felt good being with her, but her message was severe. She told me to destroy the book, to start over. It was very clear."

Leslie advised me to take my time in order to understand all aspects of the guidance. But the more time I took, the clearer the message emerged. I needed to develop a thought atom if I wanted to write. No question about it. I destroyed the manuscript, a fair exchange since I had gained Alpha Hummingbird of the hither world.

The years passed and Alpha became my sacred messenger, providing insights and highlights of wisdom from the hither world and its inhabitants. I sat in my little office and watched her lightly balance on the swaying tip of the young pine outside my window, and another secret would emerge from the ancient heart. I dreamed occasionally of jaguars and pyramids. I read books on DNA and quantum mechanics and watched Alpha rearrange Twenty Count into a series of double helixes and unidentifiable molecules. I noted that the double helix was a perfect replica of a joint double. I wondered if Twenty Count's potential combinations were hinting that additional strands of DNA might exist, that humankind's potential lies far beyond our wildest imaginings. I drew straight-lined geometric constructions encasing circles and spirals. I sketched the black diamond of the human, then erased the black in an act of vacuuming the shadow. I opened a book on the new physics and realized I had created a lattice diagram. The thought atom was present. And beyond my control.

Reflecting with Wolf in meditation on my visit with Karma Master, I remembered a series of flickering lights of guidance. They became my seven principles for developing a thought atom:

- You can seed a cosmology in the blink of an eye. Take time to prepare your seeds properly.

• The ideas imbued with most passion will shine brightest. Focus your passion where it's most appropriate.

• Cultivate rebellion. Be radical. A new cosmology requires new ideas, not recycled dogma.

• Develop space between thought and words, but not between thought and action. Speak clearly, act decisively.

• Everything has the capacity to keep expanding. It's simply a matter of where you choose to stop. If you hold back while your essence wants to grow, illness will result.

• Synchronize your thought atoms with those of the world and the cosmos. You don't live here alone.

• Release your burden of choice in order to proceed with confidence. Trust your thought atom. Trust the process.

Each guideline inspired a different focus, and together they generated some major inner excitement. All seemed a part of Karma Master's communication and all were blessed by Lama, the guide of my new history. With each thought, the scenario between Karma Master and me took on added essences of reality. The entire sequence made so much sense as Wolf helped me reconstruct it that no doubt lingered that this information accurately reflected the exchange between Karma Master and me that night on Leslie Gray's office floor.

Eventually, I also was gifted with the thought atom known as the spiraling cosmos.

THE SPIRALING COSMOS

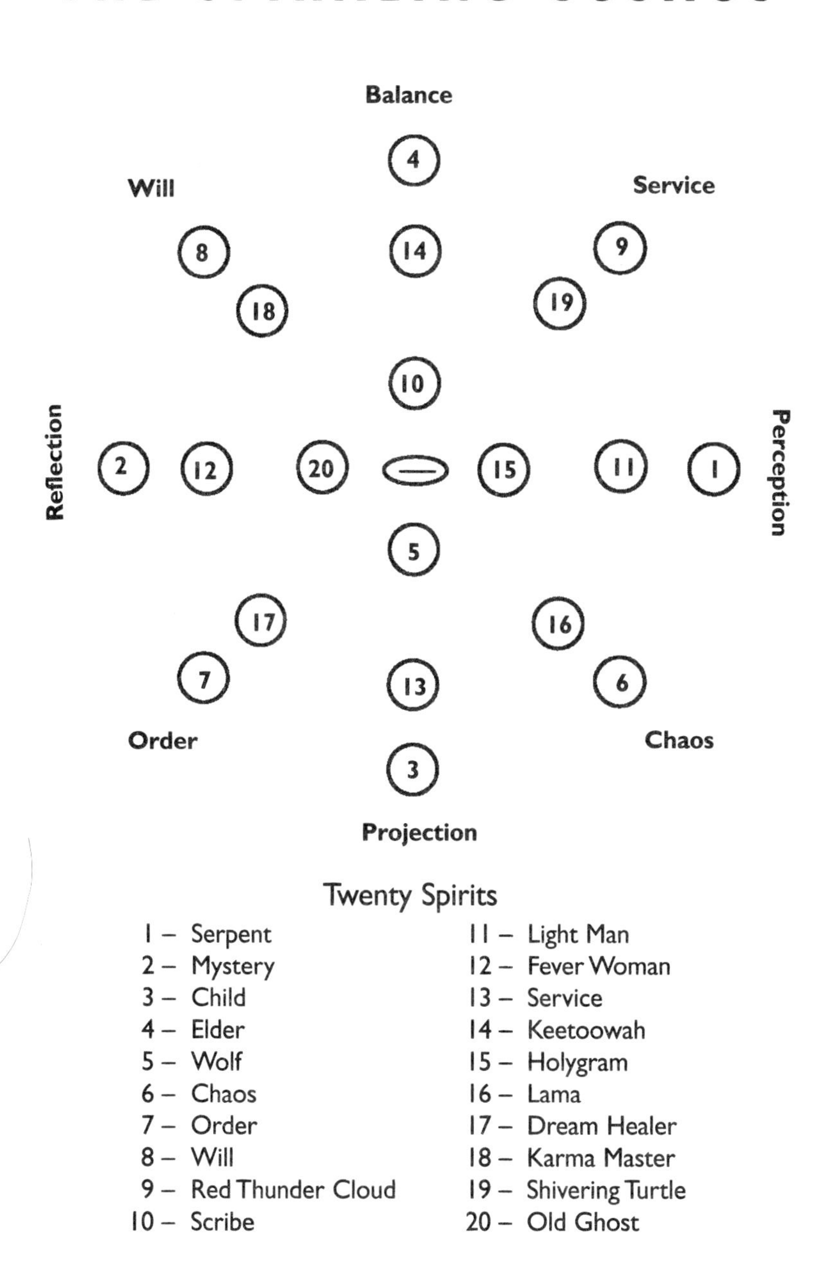

Twenty Spirits

1 – Serpent	11 – Light Man
2 – Mystery	12 – Fever Woman
3 – Child	13 – Service
4 – Elder	14 – Keetoowah
5 – Wolf	15 – Holygram
6 – Chaos	16 – Lama
7 – Order	17 – Dream Healer
8 – Will	18 – Karma Master
9 – Red Thunder Cloud	19 – Shivering Turtle
10 – Scribe	20 – Old Ghost

Shivering Turtle

"I am Shivering Turtle, the teller of the stories of arrival. Round and round all your little circles and cycles and spirals I have traveled, watching and gathering. Now I hold on to no-thing, and no-thing holds on to me. I am your gatekeeper to the passageway into the hither world. This is your life's time in which to perform your act of service. If your story of arrival is ever to make sense, it must manifest now through your spiraling cosmos. This is your thought atom."

He was tall, a traveling teller from a century far away, perhaps past, perhaps future. His deep-blue eyes focused intently and his voice resounded clearly. He smiled easily and he spoke with impeccable articulation. "I am everything you would like to see in your ideal spirit guide," he explained. "But if you wish, I can shape-shift easily. I can be anything your other spirits have been. You've arrived at a quite sophisticated level."

Alpha the Hummingbird flitted into my field of vision. I felt more at ease than in any previous meditation I could recall. Completeness was the only description for the feeling.

"You have reached a condition wherein you can travel about both your material and dream worlds and experience their blending," Shivering Turtle went on. "That's what happened in your stew at Eighteen. The best of all your realities surfaced and seasoned one another. For each breath you take, each in and out process during which you cross over and back through the ancient heart, you establish a reality. Often these realities are so fleeting that you cannot pin them down. But you certainly are familiar with what–right now–you call the real world. And you also remember the abstract realm you call Twenty Count. So, as a result, you are capable of traveling through your real, material world in full contact with the power and process of your Twenty Count."

"But your point," I stressed, "is that this is a nonstatic existence, true? Always in motion, right? But that I can experience a flowing, grounded balance, taking advantage of my sense of identity through a magical approach of rhythm and harmony in day-to-day living?"

"You're starting to talk like a spirit now," Shivering Turtle laughed. "What you're really trying to say is that you've learned to act with complete confidence. That's what you picked up at Fourteen, the power of instinctual intellect. You learned to trust your self and your body. With your magic out on the surface, you're much more interesting. Well, at least you talk better. Let's check your perception. I wonder if your seeing has improved. Take a deep breath, hold it, then breathe out. Now look around, experience the surrounding phenomena as directly as possible, and record within your memory what you perceive."

In one instant the spiraling cosmos manifested, phenomenal yet unlimited, universally comprehensive yet personally specific. The ancient Zero of the Maya scribes was the hub of the universe, the Milky Way spiral. By concentrating, I could make out my own guiding Scribe penciling the image of the spiraling cosmos onto my field of vision. Surrounding that Zero, the circles of Twenty Count vibrated all together, no chronology in sight. I focused on east and read the circle clockwise: perception, chaos, projection, order, reflection, will, balance, and service. This was the world's Twenty Count, an aligned flow of wind that encompassed and integrated both concepts and processes.

If I could align myself with this easy flow of integration that began with perception and ended with service–or more accurately, led to service that enhanced the next perception–I would be traveling neither fantasy road nor coyote road. By perceiving chaos and projecting order through the cleansing and creative process of reflection, I would create my own will and balance and service. This was Earth's Twenty Count, a process of successful flow if I could but tap into it.

"You're right," Shivering Turtle said. "This is the magic that frees mind and body from being stuck in the structures and pains of being human. You can apply it to the streets and stores and stress. This is creating your own business rather than being bound forever in the nine-to-five workplace. This is releasing your guilt and anger toward your family. This is learning to love your life in every moment."

As I peered deeper into the memory folds of the spiraling entities, the cosmos' Twenty Count also pierced my awareness. All the wheels of Twenty Count dissolved into twenty sub-wheels, each with an ancient Zero as the center. Scribe had created them all. Every single sub-wheel beckoned me, and I knew a spirit lived within each.

"Yes," said Shivering Turtle, "those are your possibilities. You thought there were only Twenty Spirits to meet. But in truth, there are four hundred. Well, to really speak the truth, I must tell you that each of those four hundred has twenty more sub-sub-wheels. They're all holographic. And beyond that . . . well, do you get the idea?"

"I get the idea," I mumbled. "I get the idea there's no end to this expanse. The cosmos goes on and on, and my only choice is to keep opening the creative center in front of my eyes. But by doing so, I will continue to come into contact with a perpetually expanding universe."

"Your field of view is the hither world," the spirit said. "You have already met one inhabitant, Dream Healer's little Alpha Hummingbird. If you scan your deep memory, you may find you've met others as well. It's true that by looking outward you perceive a continually expanding universe, but there is a tremendous advantage in reaching this particular vantage point. By turning your head ninety degrees to another angle, you will shift your vision of your own Twenty Count's rotation within the world's Twenty Count.

"Think of it this way. When you initiate perception on fantasy road, where you start with perception and then cross the circle to reflection, then go south to projection and north to balance, you create a cross. And remember–a cross with no circle creates pain with no power. So where do you get the circle? By visiting your history, dreams, relationships, and choices? By realizing your expanded awareness and then accessing kaleidoscopia to manifest remembrance, magic, alpha empowerment, and the freedom from choice? Well, it's a wonderful spiral path, but it doesn't provide contextual meaning for your existence.

"That outer circle is the world's Twenty Count: the simple, flowing, clockwise rotation beginning with perception, flowing around the wheel and back again to renewed perception. In other words, your personal process of awakening must reach realization in the world, in terms of service to Mother Earth, in order to become an effective and beneficial process for either you or the world. Your personal process is the cross that must become centered within the wheel of the world. This centering creates a complete thought atom. You are part of the whole, you're not here alone. That is the medicine wheel, and its meaning is very clear."

Shivering Turtle explained that the most efficient path through all this complexity was also the longest and yet most natural road for Serpent, the path that winds around and around and keeps going until it returns to the source. This, he said, is the true focus of Twenty Count: the journey out of the personality of separated self, through expanded awareness into kaleidoscopia, and across to the hither world. Landmarks along the journey are the dissolution of the mind barrier, the resolution of the mystery game, the mastery to enter and leave the ancient heart at will, and the personal freedom awakened by remembering that you have forgotten the shine of the source.

"The mystery of the medicine wheel reflects that very shine from the source," the spirit explained. "The deep memories of native cultures throughout the world echo Twenty Count's teachings in culturally stylized stories. Ancient myths common among those who remember tribal tales account for our presence on Earth. We have come here, say the myths, from other planets beyond our current range of vision in order to seek a better home or, in some cases, to further the well-being of our far-away source among the stars."

It has been suggested in the writings of Zecharia Sitchin, a scholar of ancient civilizations, that habitation of Earth might be part of a cyclical process that spans millions of years and immeasurable distances. This theory projects repeating cycles of time during which humans destroy their home planet's ecology, transport to another in spaceships, and populate a new colony as the old planet dies. The memory of this relocation persists in only a few minds amid a mass population lacking more than a scant grasp of historical perspective. The colonists eventually develop a science-based culture that inevitably destroys the new planet while progressing in such areas as, say, space travel, enabling still another planet to be colonized. The scenario repeats again and again, until it is impressed deeply into our magic memory, perhaps to emerge as science fiction stories about a spiraling process of mass amnesia. Unless the deeply concealed thought atoms are recovered, the true history will be lost forever.

Correspondingly, we see the same process within each human being: the entry of a spiritual intellect into the physical body that wears out just as it awakens to its own knowledge and develops the wisdom to proceed into the next realm at death. We condition ourselves to fear the return to our source because we instinctively know that destruction is lurking there. Our perception of self-knowledge has evolved as a mythological monster lurking at the base of the spine, something secret, theoretically perceivable only by gurus and fundamentalist ministers. Our fantasies stress that the pursuit of knowledge is a dangerous undertaking. The antidote to that fear exists in the secrets deposited deep within our minds, and the spiritual trip is our attempt to access those secrets.

"The spiritual urge for this antidote to fear generates the force that makes the tumblers tumble as hidden agendas in search of wholeness," Shivering Turtle continued. "It is the wind that whirls vibrations into alignment from the ancient heart on the quest to expose buried thought atoms. Each aspect of creation is seeking its complement in order to restructure completeness. In mathematical terms, each of twenty numbers is seeking its missing aspect in order to reach Twenty, its doorway back into the creative center where it can regenerate and project its awareness. Two is seeking Eighteen, Five is seeking Fifteen, and so on. Twenty is the goal, the unified thought atom."

"So if, for example, Five is seeking Fifteen in order to reach Twenty," I said, "that means that separated self is searching for kaleidoscopia, correct?"

"Right," Shivering Turtle answered. "It means that you must know yourself fully in order to understand how to live up to your potential."

"But how can I know what number I am? You've not given me a formula to figure it out. And if I don't know my number, how can I know the number of the goal that I must seek in order to reach my potential?"

"I thought you had left all this behind back at the junction of chronology and synchronicity. Your purpose is to know yourself, not your number. Part of you is always present in every number. Just remember, the full context and meaning of any experience, including your spiritual trip, becomes evident only in retrospect. Only by taking a backward view can you gain the perspective necessary to appreciate the path you have walked. It doesn't matter where you are on your path in terms of numbers or definitions.

Just know that if you expand your mind, the next step will offer itself. After all, if you're at Fifteen, you may not want to add Five to reach Twenty. It might be better to add five more Ones, or maybe Three and Two. The obvious step is not always the best alternative. That's why I like my name. Turtle sees the alternatives available through deliberate movement."

"So the answer is instinctual intellect?" I observed. "You just always have to know who and where you are."

"It's not like you don't have guidelines along the way. Instinct is not the final answer to every question. Your body can tell you many things if you learn to listen to it. Synchronous events produce inner sensations. If your focus shifts from your own Twenty Count to the world's and you end up aligned with a vision too intense to handle, you'll be uncomfortable. Your body lets you know when you enter new territory. You take in the energy of a larger environment, which is completely invisible. You experience anxiety, head-throb, heart palpitations, nausea, and paralysis that can't be explained. You and Mother Earth have a holographic relationship. Your body feels peaceful when the larger body is at peace. That's why you feel good out in nature. But your little body also feels it when Mother's body goes through an earthquake, her own version of baby death. It's a matter of synchronizing your own personal flow with the larger flow of Mother, and secret synchronization guidelines are hidden everywhere.

"It's like when you meditate. In order to see the universe, you must be able to step back so that you're not blocking your own view. Just as your heart beats in and out, and the breath flows in and out, so do you travel back and forth between your own reality and the hither world. Contradictions resolve through breathing synchronously in harmony and rhythm with the cosmos. The length and intensity of breath corresponds to the distance and direction you perceive as path. Its movement determines your larger environment–you must keep the breath flowing in order to continue the spiral perception within. If you can keep the breath in constant motion, there is no limit to the spiral possibilities. Eventually, if you develop the physical grounding to maintain your posture and follow your breath line perfectly, you can affect your environment in fascinating ways. You might even defy gravity. There are realms where it's not a law.

"On the other hand, if you hold your breath, the movement becomes linear and your perception flattens out into fantasy. And humans certainly like to stop and hold their breath. That's like wanting to identify with just one position of Twenty Count forever. If you access certain spirits in the sub-wheels of Twenty Count, they will teach you to practice your own dying by breathing in a controlled flow right through the change of worlds. If you hold your breath at the moment of death, all your worst fears will manifest and you will miss the incomparable beauty of the lights and colors and sounds of crossing into the hither world."

Concentrating on my breath, I closed my eyes and Lama appeared, jolting me to an inner-vision doubletake. I could have sworn that, during the split-second my image needed to focus itself into clarity, Lama had appeared in a Cherokee turban. He held out a closed fist as though to deliver a gift, and I instinctively reached toward him. But before we could touch, he opened his hand to reveal the tiny carved crystal skull I had last seen disappearing off the pier at Avila Beach. Peering into the depths of the crystal, I heard the ocean roll and roar.

"The future is in the crystal," Lama said. "I see you going home, back to east."

"Back East? Hey, I can go see Red Thunder Cloud again. I haven't been to New England in years."

"Not your east," Lama said. "My east. Take your woman and hit the Himalayas."

My childhood dream. The Himalayas, Potala Palace, monks in saffron robes, the abominable snowman. I liked his view of my future.

"I also predict," Lama added in a perfect imitation of a side-show fortune-teller, "that you will find something you've been seeking for a long time. Something tangible, something that someone else asked you to find. Something that maybe you've forgotten."

Instantly, Lama evolved into the persona of Holygram the hippie. "If you want to see where you're headed, man," Holygram laughed, "you need these." He projected a gift to me, his hallucinogenic sunglasses. "Be cool. Put them on and gaze within. What do you see?"

Light Man and Fever Woman appeared on cue, passionately engaged in wild and wonderful lovemaking. Both beings blended into their heat, loving and copulating with an inexhaustive driving force. Right before my eyes, they gracefully maneuvered into a new and apparently even more exciting position, and then again on my left, and my right, behind me, above me, below me, within me. For an instant, they looked exactly like dolphins. My perception swam in their force and flow.

"It's going on everywhere, man," Holygram laughed. "May as well get used to it. You may think that teaching is good and telling is cool. But I'm here to tell you that touching is downright great. Feel that energy? That movement that creates and inundates every undulation. That's love, man, real love. People think love's some syndrome of socially approved pairing off with courtship and heartbreak and sacrifice. Bull, man, bull. Look at what you're seeing. That's love, and it's everywhere."

From my current viewpoint, love looked pretty good.

"Well, maybe it's time for you to relax a little bit." This voice was Wolf. "Quite a heated pool you've been dipping into. Do you have even the vaguest idea where you are?"

"In the cosmos, I presume," I said. "Or maybe in the ancient heart. Or lost somewhere in the mystery game. What do you think?"

"I think it has been a long count, but you are now Master of Twenty Count."

Wolf's pronouncement evoked a strange effect in my inner mind. Master? Me? Master of Twenty Count. After all this time, I've finally reached the goal? I've made it? Me? I?

"Wait a minute, Wolf. This is another 'You're the one,' right?"

A coyote chorus howled in glee. I realized then there was no such thing as Master of Twenty Count, that the pursuit of the mystery game was indeed an eternal spiral that would go on forever. Neither supreme goal nor ultimate reward lurked at the rainbow peak of the path.

"No goal? That's not exactly true," Wolf said. "You've forgotten something, and you're still not remembering it. As Old Ghost told you, there is an act of service to perform, an act you've started, but haven't completed. You have created the basis for the act by perceiving a thought atom. You remember? It was done in steps. First, you mastered–yes, mastered–your own Twenty Count by advancing into perception of the hither world, the cosmos. Then you completed your clockwise mastery of the world's Twenty Count through the process of perception of chaos, projection of order, reflection on will, and balance through service. That created the thought atom of balancing the four-directional cross of personal existence within the spiral of worldly existence. Your thought atom is Twenty Count. It is also your story of arrival, which must now make sense, just as Shivering Turtle explained."

"Maybe it must make sense," I said slowly. "But if so, I can't see it yet. Can you say some more?"

"Your story of arrival will be your act of service," Wolf explained. "You'll recognize your story by its beginning. You will recognize the first words from the source, words floating back to you from the records of your guiding Scribe. Trust me, you'll know. You'll just know."

VISION GATEWAY

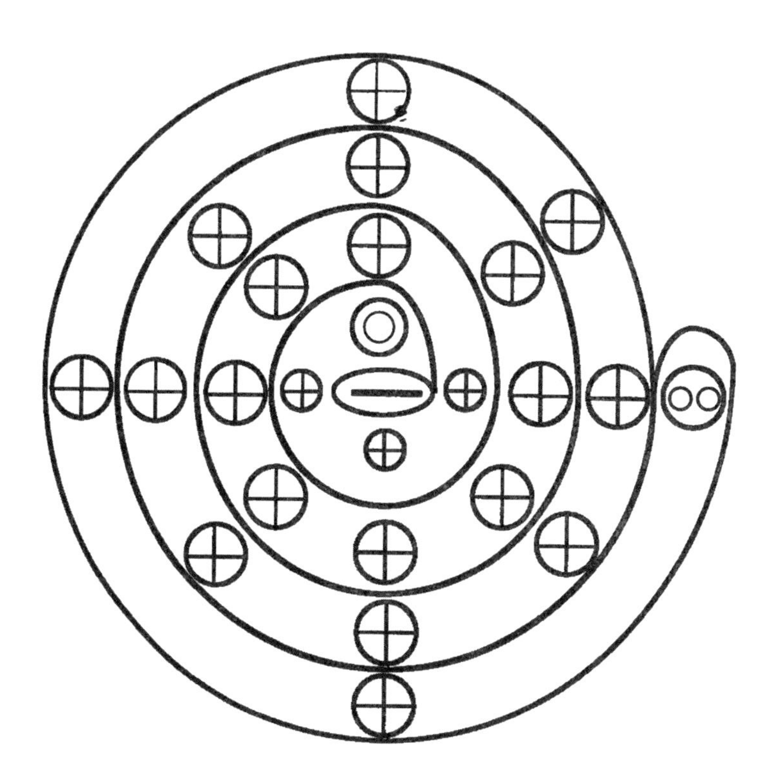

Is Zero the head?

Is duality the tail?

Free Like Wolf

The incredible Himalayan Annapurna mountain range, one of the planet's most spectacular sites, towered over us like gigantic crystals in the gleam of early sunlight. The town of Pokhara, Nepal, lay chilled and quiet this December morning. Haleh and I finished our cheese sandwiches and steaming hot chocolate just as layers of clouds suddenly dematerialized to display the spectacular, unclimbed spire of Machhapuchare rising breathtakingly above Lake Phewa.

We had just returned from a grueling trip into Tibet where we'd walked the thin air of the highest paths on Earth, and our tears had welled up amid the bittersweet burning of butter lamps in Tibetan monasteries. The Chinese invasion of the past half-century has choked this land into an oppressive mimicry of itself. The conquerors destroyed many monasteries and uncounted monks and other Tibetans have perished. Some monasteries remain, populated by a handful of surviving monks, ostensibly to service local worshippers but in reality for the benefit of foreign tourists. The butter lamps wisp their smoke toward the heavens, and age-old statues of buddhas and lamas stare blessings at visitors. But the spirit within these walls is bitter and resentful. Tibet's monasteries, esteemed in the Western world's fantasy as a source of ultimate wisdom, have been transformed into just another roadside attraction.

In Lhasa, Haleh was suffering from a combination of flu, stomach upset, and high-altitude reaction. We located a traditional Tibetan doctor, a burly, kindly grandfather who held her hand, felt her pulse, recited (through an interpreter) her medical history with startling accuracy, and gave her herbs that had her back on her feet the next morning. Obviously, some of the secrets still survive. The old herbalist was fully remindful of someone brewing a stew in a matrix.

Outside Lhasa, where the mighty Himalayas begin, Tibetan children begged for pens, cigarettes, and money. The rugged and desolate landscape is eerily remindful of the Andes range of South America. Further on, nomadic people camped with their yaks, tended barley, and persevered in the tradition of centuries. People laugh and chat, but they are careful, wary of strangers. The traveler always retains the oppressive feeling of a land under domination.

Traveling south to Nepal, we attended an exiled high lama's lecture to a crowded roomful of Westerners at the Tibetan meditation center in Katmandu. Trekkers from around the world expectantly sat cramped together for hours as he delivered a virtually incomprehensible message, tirelessly repeating barely discernible English phrases. Perhaps he transmitted something invisible to us all.

We watched a cremation by the river below our lodging at

Hotel Vajra in Katmandu, a city of great temples and global fascination and noise and filth and poverty. Haleh was working on the perception of death, and she was fascinated to see chopped-up legs and arms burning, the inner spirit obviously departed. The ashes were destined for the nearby river, already flowing with other refuse past vegetable sellers washing their trade goods.

"Life is nothing more than a preparation for death," she observed. "We get quick glimpses here of the light—the same light of passage through death. That's what it's all about, isn't it? Earth is a place of transition, and life gives us view after view of the other side, but we can't see it. We get locked in the body and blinded by everything else in the material world. The light of clarity of synchronous service is there all the time, but we refuse to see. This is another thought atom, maybe the most basic one of all."

The Hotel Vajra librarian was a seventy-six-year-old Indian yogi of impressive girth and flowing beard. He offered us paintings created by village women, preserving an ancient art form that he taught. His young American student stopped by and addressed him as "Baba." He said he could walk through the icy mountain passes in the dead of winter without a coat, using breathing discipline to retain body warmth. We did not doubt him for an instant.

All these sights, sounds, smells, and feelings created our personal lost horizon. In Pokhara, we kicked back to rest for a while at the Hotel Mona Lisa, a comfy little lodging above Lake Phewa where we could watch the mountains burst out of clouds. The night before, I sat up late in the darkness and sketched a Twenty Count spiral, the Maya Zero at the center and the primordial wheel with two inner wheels of duality at the inspirational point of east. The spiral took the form of Serpent, and either Zero or duality could be seen as beginning or end. The twenty spirits appeared as twenty spots on Serpent's body. From whatever point I gazed, Serpent was a spiral gateway to the other side, to freedom.

"It's the clearest view you have ever achieved," Wolf confided. "But you still can't tell which end is up. You're confused as to whether Zero is the head or duality is the tail. But there is a difference these days. Now, you no longer care. The desire for clarity no longer haunts you. You are freer now, a free soul like your Wolf. Like me. Now that you're here, doesn't it seem so simple, so easy?"

In the brightly emerging morning at the base of Machhapuchare, we finished our sandwiches and chocolate in the sun's warmth. I opened a copy of the *International Herald Tribune*, which had arrived in Pokhara within only a week of publication. We had been away from the mass media's incessant roar for some time, and thought we'd check the outer world. With an easy detachment, I briefly scanned the international politics that dominated the front section. The center page was devoted to movie and book reviews from New York and Paris. At the bottom of that page, my eyes fastened onto a review of *The Future*, a new album of songs by Canadian poet-singer Leonard Cohen. At the end of the article, the reviewer quoted from a song called "Anthem", which Cohen had worked on for several years, drawing inspiration from Cabalistic sources, especially sixteenth-century Rabbi Isaac Luria. As I read the four simple lines, years of thoughts and feelings aligned within me:

Ring the bells that still can ring.
Forget your perfect offering.
There is a crack in everything.
That's how the light gets in.

There is a crack, a crack in everything. That's how the light gets in.

I saw it. I heard the bells. I gave up my desire to sacrifice, my demanding need to surrender, to perform the ceremony of perfect offering. Pursuing the mystery for years across continents and oceans had been worthwhile, but it was also fine to just let it go. The twenty spirits of Twenty Count, undoubtedly far older than the ancient Maya, had guided me from Indiana through New England and California; on side trips to Mesoamerica, South America, and Europe; then halfway around the world while returning from Tibet with a woman from Iran to this simple revelation on a hotel roof in Nepal, through a Paris-datelined article quoting a Canadian poet inspired by a sixteenth-century rabbi. I had been spinning and spinning, and here was the ancient heart.

The vision was spiral, every point and spirit connected to all others, and we were simply here watching the mountains soar and learning about death. Nothing existed so solid in its core that it would not crack, and the light was ever-present, ready to flow right in. The twenty spirits of Serpent's coil were equally at home in Red Thunder Cloud's herbs, my bag of arrowheads, Keetoowah's crystals, Katmandu's cremation, Tibet's monasteries, New Mexico's sunsets, and Leonard Cohen's poetry pen. There was a crack in every shadow, a break in every pattern, and the light flowed right on in.

After rejecting the futility of pursuing clarity, I had shifted my allegiance to the perception that the highest path was the quest for freedom. Now even freedom revealed itself as fantasy midst the flow of light into shadow. *Forget any goal, simply be the process.* As Wolf once explained, I am not separated from my whole, I am just naturally complete. Separation and alienation are the true fantasies.

Observation, meditation, and poetry all confirm there is no magic answer or path, other than to just keep on perceiving the silent Oneness spiraling, ever spiraling, right in front of the eyes. We can follow it. Anyone can. And life opens and makes sense in its own time.

In the quiet morning under the Himalayas, it was just that simple. Just that easy. All I had to do was remember.

TWENTY COUNT

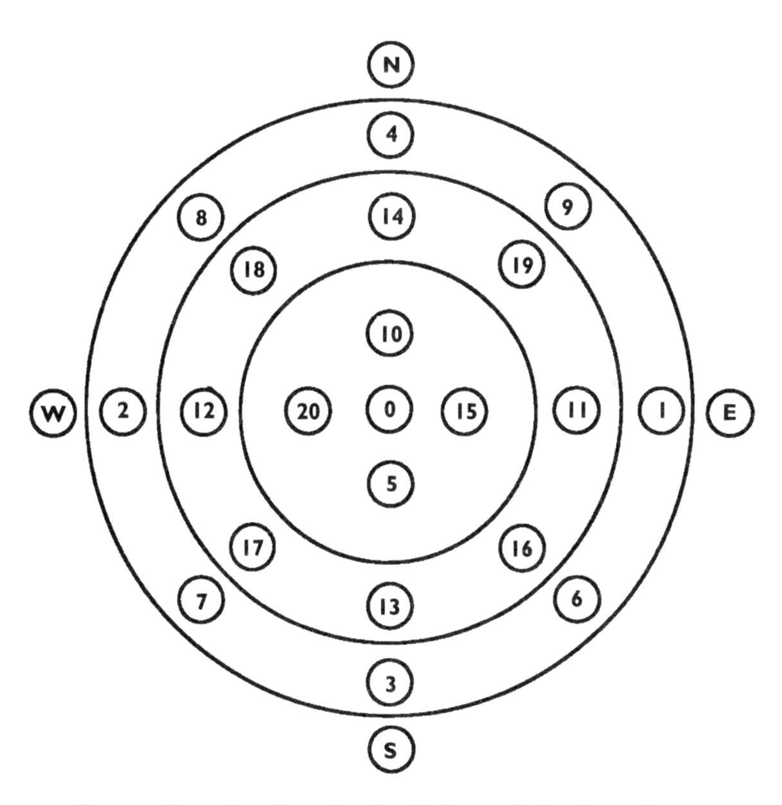

A paradigm for flow in the light and shadow show.

1 – Perception
2 – Reflection
3 – Projection
4 – Balance
5 – Separated Self
6 – History
7 – Dreams
8 – Relationships
9 – Choices
10 – Expanded Awaremess

11 – Inspiration
12 – Organization
13 – Transformation
14 – Instinctual Intellect
15 – Kaleidoscopia
16 – Synchronicity
17 – Imagination
18 – Empowerment
19 – Freedom
20 – Return to Ancient Heart

Fireflies

The priests of the cross with no circle burned the beaten-bark library long ago, and the old scribes' stories now flit among us like fireflies on a summer night. Sometimes we catch their gleam, other times we are blind to their flashing. Teachings resurface slowly with assistance from the ancients, and their light nourishes the planet. But unimaginable insights remain still unassimilated in our unified perception, and we persevere in the contradictory shadows of our human corners. We sit in the midst of this light and shadow show and see darkness in all directions.

Gaia's peacekeepers and stewards are struggling. The world's refugees are estimated deep into eight figures, and world population continues to grow at staggering rates. International developers oppose efforts to save fragile tropical rain forests, their tribal inhabitants, and their vanishing species. Native resistance leaders are targets for murder.

Destruction of tropical forests is now exceeded by loss of

temperate forests extending far north of the equator, and bitter battles rage over rights to cut old-growth timber in the Pacific Northwest. Oceanologists report that coral reefs and delicate ecosystems along polar ice and the deep sea floor are dying, threatening the oceans' balance of life. Two-thirds of all bird species are declining, and 1000 species are threatened with extinction.

Throughout the world, violence rages. Ethnic wars, gangs, poverty, and debt crises dominate our headlines. Relentlessly, Mesoamerica's and South America's centuries-old legacies of struggle continue. More than 100,000 Maya have been killed during civil war in Guatemala, and another 40,000 have disappeared. The Colombian government has acknowledged that police, judges, and soldiers allowed paramilitary groups to murder 107 people between 1988 and 1991. In Rio de Janeiro, gunmen identified as city police murdered seven street children, ages eight to twelve, on a July night in 1993. A Brazilian government inquiry reported 4600 street children had been killed in similar fashion over three years.

"Dear scribes," we are prompted to shout in perplexed anguish. "We are trying, but it is obviously not enough. What more can we do? Send us something besides fireflies."

Through it all, an ancient beneficence calmly smiles and lifts a wrinkled hand to spin a heavenly spiral beyond and above these concerns.

"Just breathe," says Old Ghost. "Accept and create wonderful solutions for your realities. Dismiss your fantasies of personality, for they will only mislead you. Revive your deep memory of imagination and thrive within your creative nurturance. Moment after moment, remember the appropriate forgetting, and observe the vision of the ancient heart. The spiral flows on and on. You've barely begun to perceive, and there is so much more beyond. Every answer resides within. When you truly get it, you'll have no question how to act in the outer world. You will be a grounded, effective, innovative human being, always able to meet the demands for increased creativity in the outer world. You'll simply observe yourself acting appropriately and confidently in all instances. You'll trust yourself without thinking about it."

Day after day, Old Ghost reveals fleeting glimpses of the mysterious range and magic of Twenty Count. The cosmos whirls forth as each of us travels our own path homeward midst a great spiral of many smaller spirals. Me, I simply breathe and watch the center open, and Wolf peers in often to remind me to walk my talk. Sometimes I reflect and feel comfortably close to the light. It even occurs to me that I understand all I've experienced. That's when I know it's dreamtime.

A small child with mischievous eyes and beaming smile approaches, takes my hand, and silently sings: Please, tell me a story. Please.

I'm not sure I can remember a story, I think-sing. Then I kneel down and take the child to my breast. Some deeply buried memory stirs. Maybe, Child, maybe you can help me remember. What is the story you want to hear?

Just remember, demands Child. That's your job. Come on. It's time for my story. Now. Now.

I breathe, and wait, and embrace Child more closely. The edges of my world begin to dissolve, and suddenly a flash bursts out of my heart. Old ghostly faces long known laugh down on me. Thunder clouds part, shadows shift, songs ring out, wheels dance and dart through the heavens. Child-life summons me. Now I can tell the story.

Like a chanter of magic memories, I mouth the message of Hummingbird of the hither world. Excited, Child claps and laughs as I commence the recitation of the ancient, well-remembered work of the scribes. The feeling is achingly familiar. So many times I've heard this same beloved story. So many times I've told it. And the first words are always the same:

Watch, my friend, and see a wondrous sight, a spinning rainbow of twenty circles of light bursting out of the fleeting gleam between elder and child, from passionate lovers one to another, from Earth to her myriad creatures. Revelations of soul wisdom dance and sing in the blinding primordial flash between birth and death. So long as you breathe on this planet, you will quest for an ancient truth moment after moment, even though I tell you that truth this very day. You see, you will not remember these words, and therefore you must always seek to hear:

"The source of Twenty Count is the singular light of soul."

ACKNOWLEDGMENTS

My thanks to Barbara and Gerry Clow, copublishers and editors at Bear & Company, for their partnership in the flow; also, to Martin Brennan, author of *Stones of Time* and *The Hidden Maya* for his rendering of the Aztec Sunstone.

And "Thank you, Bear," says Wolf.

Quotations within the text are taken from the following sources:

Introduction

Carl G. Jung, *Man and His Symbols*. New York: Doubleday, 1964; p. 14.

Dr. E.C. Krupp, *Beyond The Blue Horizon*. New York: HarperCollins, 1991; p. 4.

The Source

Kenneth Meadows, *Shamanic Experience*. Rockport, ME: Element, Inc., 1991; p. 148.

Bill Wahlberg, *Star Warrior: The Story of SwiftDeer.* Santa Fe: Bear & Company, 1993; p. 242.

Long Count to Twenty

Octavio Paz, *The Labyrinth of Solitude*. New York: Grove, 1961; p. 212.

Machinery for Change

Leonard Cohen, "Democracy." Copyright 1992 by Leonard Cohen Stranger Music, Inc. Used by permission. All rights reserved. (From Leonard Cohen, *Stranger Music: Selected Poetry and Songs*. New York: Pantheon Books, 1993; p. 367.)

Matthew Fox and Brian Swimme, *Manifesto! For A Global Civilization*. Santa Fe: Bear & Company, 1982; pp. 27-28.

In Finite Chaos

Leonard Cohen, "Magic Is Alive" from *Beautiful Losers*.

Joint Double

Leonard Cohen, "The Captain." (*From Stranger Music: Selected Poetry and Songs*, op. cit.; p. 341.)

Free Like Wolf

Leonard Cohen, "Anthem." (From *Stranger Music: Selected Poetry and Songs*, op. cit.; p. 373.)

The paradigm wheels, numerical relationships, and teaching applications in this book are original. They have been derived and developed by the author, and are not part of Twenty Count formulations developed or taught by any other individual or group.

About the Author

Roger Montgomery works with individuals and groups committed to expanding the creative vision of contemporary culture through Earth-based spirituality, visionary science, art, and philosophy. He lives in California.